T0249552

Motorcycle Tuning: Chassis

by the same author

Motorcycle Service and Set-up Data
Motorcycle Tuning: Four Stroke
Motorcycle Tuning: Two Stroke

Motorcycle Tuning: Chassis

Second edition

John Robinson

OXFORD AUCKLAND BOSTON JOHANNESBURG MELBOURNE NEW DELHI

Butterworth-Heinemann
Linacre House, Jordan Hill, Oxford OX2 8DP
225 Wildwood Avenue, Woburn, MA 01801-2041 USA
A division of Reed Educational and Professional Publishing Ltd

℞ A member of the Reed Elsevier plc group

First published 1990
Reprinted 1992
Second edition 1994
Reprinted 1995, 1996, 1997, 2000, 2001

© John Robinson 1990

All rights reserved. No part of this publication
may be reproduced in any material form (including
photocopying or storing in any medium by electronic
means and whether or not transiently or incidentally
to some other use of this publication) without the
written permission of the copyright holder except in
accordance with the provisions of the Copyright,
Designs and Patents Act 1988 or under the terms of a
licence issued by the Copyright Licensing Agency Ltd,
90 Tottenham Court Road, London, England W1P 0LP.
Applications for the copyright holder's written permission
to reproduce any part of this publication should be addressed
to the publishers.

British Library Cataloguing in Publication Data
Robinson, John
Motorcycle tuning: chassis
1. Motorcycles. Chassis. Tuning
I. Title
629.2′4

ISBN 0 7506 1840 X

FOR EVERY TITLE THAT WE PUBLISH, BUTTERWORTH-HEINEMANN
WILL PAY FOR BTCV TO PLANT AND CARE FOR A TREE.

Transferred to digital printing 2006
Printed and bound by Antony Rowe Ltd, Eastbourne

Contents

Preface

Racing, the whole fun of riding bikes and the practicality of riding bikes, is all about going around corners. A machine which can corner faster or which is more manoeuvrable, will be more competitive, more fun, or easier (safer) to get through traffic.

Anyone who has ridden more than three bikes will know that there are differences – sometimes big differences – in the way they react. Anyone who has tinkered with suspension and tyres will know that it is possible to make big differences with small changes. And that other changes (usually the adjustment provided by the manufacturer) make practically no difference at all.

It is a subject which is difficult to quantify. Some people have a natural feeling for it, some do not. When I understand how a part works and when I can visualize what is happening to it, then I can work out how to adjust it. In my case it is not a natural feel, in fact it is usually the opposite, a long drawn-out sequence of trial and, mostly, error. It teaches one thing: believe the stopwatch.

What I have tried to do is set out a logical sequence to build a new bike, or to modify an existing one, to make the most use of its power. Some things can be quantified. The geometry of the bike dictates its ability to accelerate and to brake: these things can be measured. The quality of the tyres controls the amount of force which can be transmitted to the ground. Beyond this, things become more qualitative and must be assessed rather than measured, which is why serious race teams spend more time testing than they do racing. If this book saves you half a day's testing, it will have paid for itself.

John Robinson.

Acknowledgements

My thanks to Colin Taylor, who wrote the chapter on welding and brazing, and to the many other individuals and organizations who provided information, illustrations, components to be photographed and were generally very helpful whenever I got stuck:

BMW (GB), Cibié, Gerry Daubney (M R Holland), François Decima (Michelin), Duckhams, Dunlop, Ferodo, Patrick Gosling, Heron Suzuki, Paul Hobbs (Goodridge), Honda UK, Kim Hull, Kent Aerospace Castings, Kawasaki Motors UK, Steve Lythgoe (Sharples Service Station), Phil Masters, MIRA, Mitsui Machinery Sales, Leon Moss (LEDAR), *Performance Bikes*, Pirelli, Ron Williams (Maxton), Yamaha nv.

Chapter 1

Steering and handling

While engine performance can be quite clearly defined and measured, the properties of handling, roadholding and ride comfort are not so easy to deal with. The sum total can usually be valued in lap times, or simply in terms of how pleasant the machine is to ride, but often it is not clear what is contributing to what. How is it that one bike is so much easier to handle than another when, superficially, they are so similar?

Much of it is subjective – which does not make it any easier to measure – and what suits one rider does not always suit another. Expressions such as 'slip', 'slide', 'oversteer', etc., often mean different things when they are used by different people. (The definitions of all these terms, as they are used in this book, are given in the Appendix.) On top of all that, the things which control the suspension and the steering are closely related; change one and you stand a good chance of changing two or three others at the same time. This manages to do away with the first rule for test work – only change one thing at a time.

To take a simple example, suppose you decided to change the rear spring. You would expect a softer or a harder spring to give a more supple or a harsher ride and so change the ride comfort. But it could also change the ride height, depending on the length of the spring and its pre-load. This would alter the ground clearance and may restrict cornering speed even though the comfort and the feel of the bike had been improved (with a view to *raising* the cornering speed). Lowering the back of the bike has the effect of raking out the front and increases the trail slightly; this makes the steering a little heavier and makes the bike more stable. Raising the back has the opposite effect, it reduces the trail and makes the steering lighter and faster to respond. For a given suspension force, the new spring will deflect by a different amount, altering the speed of the damper and therefore the damping force it generates. So just by changing the spring rate you could make noticeable changes to the way the bike handles and steers. You could also alter its straightline stability and its cornering clearance.

Obviously, any changes have to be worked out quite carefully, so that they do what you intend with no unwanted side effects.

With all of these factors – trail, stability, spring rate, etc. – there is an optimum level, which matches the rest of the machine, the rider and the road or track conditions. Where a new design, or a seriously modified machine is being developed it is necessary to work out a series of tests which will produce the best blend of chassis and suspension. On the other hand, if the

1

work is aimed at curing a specific fault, such as weaving or wheel patter, then experiments on a least effort/least cost basis will be as good as any other method. Because one change can have so many effects it is rarely possible to predict the results completely accurately and so test riding is essential. It also makes handling and stability faults very hard to trace; in fact it is often easier to try to make the fault *worse* so that the responsible part can be identified.

And when it is all finished, one rider may like the resulting 'feel' while another may not. A third rider may not care; there is a lot to be said for employing a rider who can go fast on anything, but do not employ him as a development rider.

In some cases the feel of the bike simply makes it more pleasant to ride. In other cases it can make the bike faster because it is able to turn faster, to brake harder or to accelerate earlier, etc. Although it is difficult to measure the individual contributions, the sum total can be seen in shorter lap times, reaching a higher speed at the end of the straight or being able to brake later. On a road bike it is usually enough just to make it feel better, small gains in cornering speeds do not make much difference to journey times. However, it follows that a bike which is *capable* of greater performance is safer to use at the original level of performance and will not put as much stress or fatigue on its rider.

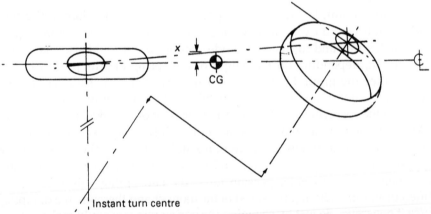

Instant turn centre

Figure 1.1 Steering the front wheel makes the machine turn about the instant centre where the axes of both wheels cross. The effect of trail moves the front contact patch, so the centreline of the bike – and the bike's centre of gravity – is no longer directly above the line along which the bike is supported. The effect of steering castor also makes the front wheel lean (exaggerated in the diagram), so it will generate camber thrust

The way in which a bike steers is the focal point of its handling and its stability. At very low speeds, a bike steers by turning the front wheel into the corner. With the bike essentially vertical, the trail (see Appendix) shifts the front tyre's contact patch to the left in a right turn. This means that the centre of gravity is now to the right of a line drawn between the two contact patches – the line on which the bike is supported. It would, therefore, try to fall over to the right.

At the same time it is moving in an arc to the right; the centre of this curve would be where lines drawn through each wheel spindle cross one another. It has acceleration (centripetal acceleration) towards this point, even though its speed, as recorded by a speedometer, is constant. The acceleration is v^2/r, where v is the speed and r is the radius of the arc which the bike is following. The force which provides this acceleration is mv^2/r, where m is the mass of the bike, and it is generated at ground level. The reaction caused by the inertia of the bike equals this force but is in the opposite direction, away from the centre of the turn (centrifugal force) and is considered to act through the bike's centre of gravity – which is well above ground level.

The centrifugal force sets up a couple which tries to make the bike fall over to the left. The strength of this couple is $y\,mv^2/r$, where y is the height of the centre of gravity. The couple trying to make the bike topple to the right is mgx, where mg is the total weight and x is the amount the centre of gravity has been displaced from the bike's line of support.

When the bike is turning in a balanced fashion it does not fall over, so these two must be equal:

$$ymv^2/r = mgx$$

or:

$$v^2 = rgx/y \qquad \text{(note that this is independent of the bike's mass)}$$

Now g is constant, and so is y for a given bike, while x depends on the steering geometry and the steer angle, and r also depends on the steer angle. For a given bike and a given steer angle, rgx/y is constant. There can be only one value for the speed v which will satisfy this and keep the bike in balance (there are two actually, $+v$ and $-v$, because the mathematics allows for you to be able to ride backwards at the same speed).

If the rider goes slower, then ymv^2/r will be too small and the bike will fall to the right, into the turn. If the rider goes faster, ymv^2/r will become too big and the bike will fall to the left, away from the turn.

As an aside, the rider can shift the centre of gravity, especially on a very light bike, by standing up (increase y) and by leaning his body to one side (increase or reduce x). So by making x and y variable, a trials rider can give himself a range of speeds for which the bike is balanced in a given turn.

If v is steadily increased and the height of the centre of gravity stays the same, then the term rx will also have to be increased. To increase x, the steering has to be turned further into the turn, but this action reduces r, the radius of the turn, so we will quickly reach a value v for which the steering geometry cannot cope. It happens, on conventional machines, at something

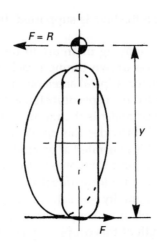

$F = R$

y

F

Figure 1.2 The cornering force F is generated at ground level, while the reaction of the bike's inertia R acts through the centre of gravity. The two are equal, so $F = R$ ($= mv^2/r$), and they try to overturn the bike to the left, with a couple Fy. The centre of gravity is also displaced a distance x from the axis through the tyre contact patches and this tries to overturn the bike to the right, with a couple mgx

in the region of 2 to 4 ft/s (1.5 to 3 mph), but it is worth remembering that this critical speed exists.

If the speed is too great – as it must be if the bike is to exceed 4 ft/s – then the effect of a right steer effort is to make the machine fall, or roll, to the left. So now we have a bike travelling at very low speed, but more than 4 ft/s, and the result of applying a right steer angle is that it rolls left. The immediate effect of this is that the centre of gravity is now displaced to the left of the line on which the bike is supported, and both wheels are leaning (or are cambered) to the left.

So far our bike would have steered and generally behaved as predicted if the wheels had simply been thick wooden discs. Now the tyres begin to do something: they generate what is known as camber thrust. Because the tyre can deform to a flat contact patch where it meets the ground it can, when it is leaning over, behave like a section of a large cone, lying on its side. If you roll such a cone it will turn in a circle which has the tip of the cone at its centre. If you take a tyre, lean it about 30° from the vertical and roll it slowly forward, it will behave in the same way.

The bike's inertia wants it to keep travelling in a straight line. The camber thrust from the banked tyres wants it to turn in quite a tight circle. The force generated is not enough for this rate of turn, but it is a force and it does make the bike turn. The force acting on the mass of the bike gives it an acceleration to its left. The value of this equals v^2/r, so the bike proceeds at whatever speed v happens to be, to turn on a radius of r.

As the bike rolls left, the reaction of the ground supporting the front tyre

4

also tends to turn the steering to the left; the rider will feel this as a reaction in the handlebar which he can either oppose, ignore or augment.

We now have a situation which is very similar to the first one: the bike is steering left, is generating cornering thrust to the left (at both wheels this time) and has displaced its centre of gravity (considerably further) from the line on which it is supported. As before, if the force trying to make it roll left (mgx) equals the centrifugal reaction ($mv^2 y/r$) then the bike will be balanced and will follow a circular path, radius r, at a constant speed v.

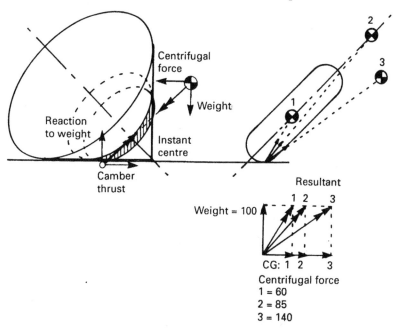

Figure 1.3 A leaning tyre tries to behave like a part of a cone rolling on its side and this sets up camber thrust. For the bike to remain balanced the resultant of the cornering force and the reaction of the bike's weight against the road must equal the resultant of the bike's weight and centrifugal inertia. Wide tyres cause the contact patch to move away from the bike's centreline which means that the height of the centre of gravity influences the relationship between the angle of lean and the cornering force. The forces necessary to satisfy three centre of gravity positions are shown. 1 Low; 2 High; 3 Shifted inwards

The displacement x is now a function of lean angle as well as steer angle ($x = y \cos \theta$, where θ is the lean angle of the bike to the horizontal) and the value mgx can satisfy a much greater range of speed and radius (v^2/r) values. An increasing lean angle also lowers the height of the centre of gravity, which reduces the tendency for the bike to roll to the outside under the influence of centrifugal force. The equation now becomes:

$mgy \cos \theta = mv^2 y \sin \theta /r$

or:

$v^2 /r = g/\tan \theta$ (this ignores any displacement owing to steer angle)

The bike's lateral acceleration becomes 1 g at an angle of lean of 45° and Table 1.1 shows the speed at which curves of various radius can be taken with this acceleration and smaller acceleration values.

Table 1.1

Angle of lean from vertical	Acceleration v^2/r (ft/s²)	(g)	Radius (ft)	Speed (ft/s)	(mph)
45	32.2	1	100	56.7	38.7
			150	69.5	47.4
			200	80.2	54.7
			500	126.9	86.5
30	18.6	0.58	100	43.1	29.4
			500	96.4	65.8
15	8.6	0.27	100	29.4	20.0
			500	65.6	44.7

This ignores any effect made by the steering being turned, which would displace the centre of gravity further still. It also ignores a couple of other effects produced by the tyres.

Because the tyres are relatively wide and more or less circular in section, the contact patches move to the left as the bike rolls to the left, that is, it is no longer supported on its centreline, as the calculations assume. The implications of this are:

1   The centre of gravity is not displaced as far as we thought it was, so the cornering force has to be reduced proportionately for a given angle of lean.
2   Bikes with a lower centre of gravity will have to lean further to achieve the same balance as those with a higher centre of gravity.
3   The wider the tyre, the worse it gets.

The second aspect has been hinted at: the tyre wants to run on one course but is forced away from it by the inertia of the bike. In addition the tyre contact patch can be regarded as part of a cone, that is, it has a greater radius on the outside edge than on the nearside edge so the outside edge will travel correspondingly faster. It is not allowed to do this, so part of the contact patch must slip (see Appendix). The tyre can also run at a small angle to the one in which it is pointing and the more it is loaded, the more it is inclined to do this. The angle is called the slip angle and if the rear tyre has a different slip angle to the front, then the bike will rotate or yaw, it will turn to one side as it travels along. If the slip angle is bigger at the rear than at the front, the bike will yaw into the turn. This is called oversteer. The rider can reduce his steering effort and still hold the same rate of turn.

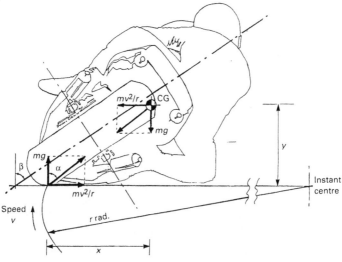

Figure 1.4 The total cornering force acting on a machine. The roll couple mgx is equalled by the centrifugal couple mv^2y/r. The effective angle of lean, α (tan $\alpha = x/y = v^2/rg$) will be less than the tyre's angle of lean (β) unless the centre of gravity is shifted further into the turn than the tyre contact patch

Depending on the tyre construction and the compound from which it is made, this slip can increase the grip available. As the slip is increased, the grip also increases, reaching a peak when the slip is a few per cent higher than the tyre's average speed. Beyond this peak the grip falls away again; when the tyre begins to slide or spin, the grip falls away more severely.

All this has been applied to a bike moving at steady speed. If the bike is also accelerating then it must be transmitting torque through the rear contact patch, and this will induce more slip at the rear tyre. This helps the cornering power because a certain amount of slip gives more tractive effort at the tyre and this can now be controlled by using engine power as well as speed and angle of lean. It is also easier to control it this way. And it is creating more slip at the rear wheel, which leads to oversteer – which accounts for the sensation of 'drifting' when cornering under power and is the reason that bikes feel more secure and controllable in this condition.

Oversteer is a natural tendency for bikes because they have rear-wheel drive. It is also the safest condition because it 'pushes' the back tyre and, if this should spin or slide, it is easier to control than a front wheel slide. Most passenger cars, on the other hand, are set up to understeer (so that the front slip angles are greater than those at the rear). This is because the steering is more predictable – more effort equals more turn – and, if the front should lose traction, backing off the power will be enough to regain control. An oversteering car requires much more skill to control and once the back wheels lose traction the car is likely to spin whether the driver applies power or takes power off.

7

When a bike turns in to a corner it takes a 'set': the speed, the rate of turn, forward or sideways shifts of the CG and the use of power (or brakes) all load the suspension, tyres and steering so that the bike adopts a particular attitude of bank, steer angle and suspension deflection. For the same bike, this can be altered significantly by the rider (for example by moving the CG, by altering the rate of roll, by the use of engine power or brakes), and can produce at least a 10% difference in cornering speed for the same angle of lean – or a similar change in lean at the same cornering speed.

This was demonstrated during some tyre tests for *Performance Bikes* when a speed trap at the apex of a second gear corner consistently showed differences in speed of around 4 mph between two riders, on the same bike (a Yamaha YZF750) with the same tyres and suspension settings. Ironically, the lighter of the two riders was leaning far enough to touch the footrest and exhaust on the floor at around 43 mph while the other rider was reaching 47 mph with the tip of the footrest occasionally grazing the track surface.

Factors which tend to increase the slip angle at a tyre are:

1   Flexible tread pattern.
2   Flexible sidewalls.
3   Flexible carcass.
4   Lower coefficient of friction.
5   Less pressure.
6   Transmitting engine or braking torque.
7   Larger section.

The grip available for cornering, braking and acceleration depends mainly on the tyre compound (and the road compound, too). The construction of the tyre gives it the ability to use soft compounds without overheating (see Chapter 3) while the construction of the bike's frame and suspension hold the wheels in the attitude which gives maximum traction. This is the requirement for good roadholding.

There is also the way in which the bike's attitude changes: the force needed, the rate at which steer effort produces roll, the combination of lean and steer angle for a particular speed and radius of curve. As the bike moves into a cornering attitude, the suspension is loaded (instead of the weight mg, it now carries the resultant of mg and the cornering force, mv^2/r; the total force on the suspension will be $m\sqrt{(g^2 + v^4/r^2)}$, and if $v^2/r = g$, then the total load will be 1.414 mg or 1.414 times the static load).

These forces and deflections at the steering and suspension mean that when the bike rolls into a turn, it takes a 'set' as it stabilizes. The pull and movement at the controls are what the rider perceives as 'feel' or 'feedback'. The speed with which it manoeuvres and the attitude or set that it adopts are what the rider regards as its handling qualities. Some combinations of these things feel better than others. Some are demonstrably better if they let the

8

bike turn in faster and stabilize sooner so that the rider can use power as early as possible.

The bike also has to handle bumps and changes of surface (see Chapter 5) and the final aspect is stability. This is only of any consequence if it affects the bike's handling properties or if the rider has to wait for it to settle down before he can use power. Competition bikes often have unstable steering but this would become tedious on a long-distance touring machine.

The idea of stability needs some explanation. An object is said to be stable if, after being deflected, it returns to its original position; if it doesn't, or if it takes an unduly long time, then it is unstable. An example, as shown in Figure 1.5, is a cone which when stood on its base is stable as long as the deflections aren't too great. Balanced on its tip, it is inherently unstable.

Things are rarely as cut and dried as this; for example, the same cone resting on its side isn't stable because it wouldn't return to its original position after being moved. But neither is it unstable in the sense that a small deflection

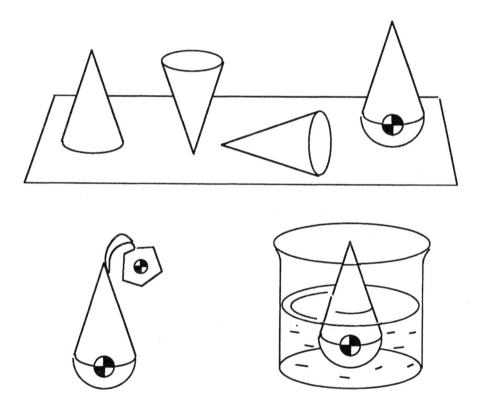

Figure 1.5 Degrees of stability. A cone on its base is stable but balanced on its tip it is unstable. A cone lying on its side is not unstable in the sense that it will fall over, but it is not stable in the sense that it will return to its original position if moved. A cone with a rounded base would swing back and forth before returning to its static position, while one with a flexibly attached mass would wobble in a less predictable manner. Immersing the cone in liquid would slow down the movement – literally damp it – and prevent the oscillation from building up momentum

would get bigger, which is the case when the cone is balanced on its tip. This condition is known as meta-stable.

If the cone had a spherically rounded base, instead of a flat one, then when it was stood upright it would still be stable but now it would take longer to return to the original position after being disturbed, and it would build up enough momentum to swing through the zero position, like a pendulum. It would swing back and forth a few times before settling and this would depend upon the curvature of the base, the height of its centre of gravity and the size of the original deflection, plus (or rather, minus) any damping forces. Damping – any force which opposes the original motion – can be illustrated by placing the cone in a beaker of water. Having to move through a more viscous fluid wouldn't alter its natural stability or instability but the speed would be reduced, preventing the cone's own inertia from creating a long series of secondary deflections. (In the case of the flat-bottomed cone the damping force could be said to reduce stability because it would increase the time taken to get back to the zero position. In the case of the round-bottomed cone it would improve stability because the cone would take fewer swings to return to the zero position.) When the oscillation reduces in size it is called convergent; increasing oscillations are called divergent.

The way in which the cone behaves depends on its shape and on how its mass is distributed – which assumes that the mass is all part of one rigid body. This need not be the case: some of the mass could be flexibly attached, in which case it behaves as a heavily damped mass, always lagging behind the motion of the main body. This is actually getting quite close to the way a bike behaves, with the rider and passenger being fairly large, heavily damped masses, but the term 'instability' has (at least) three different meanings:

1 Very light, 'twitchy' steering found on racers, sports and very light bikes, which demands constant concentration and correction even to hold a straight course. Should really be called meta-stable.
2 There are occasional, highly specific times when the steering goes unstable (the 'tank-slapper'). This is usually associated with some external disturbance such as a bump and is very difficult to reproduce on demand but there is no doubt as to the lack of stability.
3 Speed-related instability in which the steering flutters in a predictable manner. Easy to repeat.

Of these, (1) is not really instability but is comparable to the round-bottomed cone with a narrow base or a high centre of gravity and (2) is comparable to the basically stable cone with an abnormally large deflection, possibly producing enough inertia to create an instability which could be self-generating. The initial impetus would be reduced by the use of a steering damper. Item (3) is true instability, in the sense that it is self-generating and does not require a large initial deflection to trigger it off but is driven by the ordinary forces acting on the bike, just like the pendulum in a clock driven by weights or a spring.

The pendulum motion is significant. During the 1970s, Dr Geoff Roe ran a series of experiments at the Simon Engineering Laboratory at Manchester University to investigate the behaviour of castoring wheels. The broad conclusion was that even a rigid wheel (without the elasticity or camber thrusts that a tyre might provide) would go unstable at a certain critical speed and this speed depended upon the castor, trail and the inertia of the wheel about the castor axis.

This can be applied to bikes because both wheels are castoring behind the steering axis, about which they will swing like one long and one short, horizontal pendulum. In fact Roe identified what he called the 'pendulum axis' – a line drawn from the tyre's contact patch at right angles to the steer axis – and found that the critical speed depended upon the ratio of the moments of inertia above and below this line. In other words, add mass below the front of the engine and it would tend to increase the critical speed; add mass above the axis and it would tend to bring the critical speed down to a lower region. From this it seems that a top box is the worst possible place to add any weight to the machine. And as far as the front wheel is concerned, the axis is very low down, so any mass carried by the front part of the bike is bad and should be mounted as close as possible to the steering axis, to reduce its inertia about this axis. This is why brake calipers are now mounted behind the fork legs and why there is a significant improvement in steering when handlebar mirrors are removed.

Imagine the steering stem as the pivot for a pendulum; if this were clamped, horizontally, high up in the air, the bike would dangle from it – the frame and rear wheel would swing from side to side like a long pendulum while the front wheel and forks would also swing but on a much shorter axis. (The same analogy applies to any articulated vehicle or tractor and trailer unit.) The bob of a pendulum has gravity pulling downwards and the tension in the rod pulling up. When it is displaced to one side, these two forces are no longer in line and they produce a resultant force acting sideways, back to the zero position.

The further the pendulum is deflected, the greater the stabilizing force becomes. If the pendulum is released, this force will accelerate the bob back to the zero position; when it reaches it, the force will disappear but the bob will have momentum to carry it past the zero position and out the other side, where an opposite, stabilizing force will appear, getting bigger the further the bob travels from the zero line, until it brings the pendulum to a standstill and reverses its travel. At the point where the sideways force is greatest, the bob has zero speed; at the point where the force is zero (the centre position), the bob has maximum speed. Without the damping forces of air resistance and whatever friction there is at the pivot, it would carry on for ever. The period (the time taken for one complete swing) is

$$1/f = 2\pi\sqrt{(L/g)}$$

where f is the frequency
L is the length of pendulum (pivot to CG)
g is the gravitational constant.

(So if you want your long case clock to tick once per second, the escapement releases at the end of each swing, that is $f = 0.5$ and the length of the pendulum must be g/π^2 (that is $32.18/3.14159^2$ feet or 39.126 inches). Note that the frequency of the oscillation does not depend on either the weight or the amplitude of the swing, but purely on the length of the pendulum.)

If we look at a plan view of a bike travelling along a road, we can still see certain similarities, with the steering head as the pivot and the bike divided into two oscillating masses, each carrying one of the wheels. If either wheel were deflected to one side, a self-aligning torque (see Chapter 2) caused by the drag of the road on the tyre, will push the wheel back into line. The greater the deflection, the greater the force. The pendulum's forces produced by the weight of the bob and the tension in the rod have been replaced by the horizontal forces of drag at the tyre and engine thrust at the steering head – which is pushed forward, with both wheels trailing behind it.

The bike is a little more complex because the steer axis is not vertical; when the contact patch of either wheel is deflected to, say, the left, then the wheel also leans to the right, setting up a (stabilizing) camber thrust in that direction.

The bike is free in roll, so a left deflection of the rear wheel, the equivalent of steering left, will tend to produce roll to the right (see above). A left deflection of the front tyre's contact patch is actually the same as steering right, which will tend to produce roll to the left. The consequences, as all learner riders must have noticed, is that a slight imbalance causes a steering deflection which

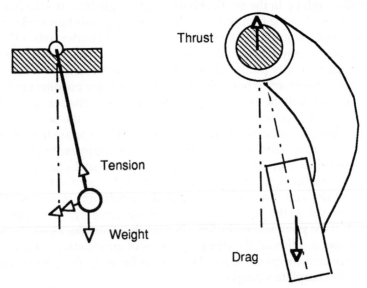

Figure 1.6 Comparison of a pendulum with a castoring wheel. The pendulum's bob has two forces acting on it, its weight and the tension in the pendulum's rod, so that when it is displaced to one side the resultant always acts towards the neutral position and gets larger as the deflection gets bigger. A castoring wheel has similar forces although they are in the horizontal, not the vertical, plane and are caused by the thrust through the castor's steering axis and the drag of the surface on the wheel, which is at a minimum when the wheel is directly in line with the thrust force

makes the bike follow a series of S-shaped weaves as it alternately steers one way and rolls the other before coming back to a straight and level course. This is more pronounced at low speed because as speed increases the bike has more inertia and the gyroscopic inertia at the wheels also increases*.

It would be a daunting job to use these forces plus the damping effect of the tyres against the road surface to try to predict a bike's likely stability or otherwise. However, we can possibly gain a little more from the pendulum analogy. It was driven by gravity, so the force automatically went up as the mass of the bob increased. This doesn't apply to the bike but we could add a term for mass in the equation for a pendulum's frequency:

$$1/f = 2\pi \sqrt{(mL/mg)}$$

This makes no difference to the equation, but it does imply that the frequency (squared) is proportional to mg (which is mass times acceleration, i.e. a force) and is inversely proportional to mass and length. The force in this case is engine thrust and road drag and, unlike the force in the pendulum which is purely proportional to the mass to be moved, the force on the bike can vary through a wide range; it can even have different values at one road speed, depending on how much the machine is accelerating. The frequency is also inversely proportional to the mass and to the length – i.e. the distance from the steer axis to the CG of the rear or front wheel assembly.

Having a relatively high mass and long length, the rear part of the bike will tend to have a low frequency, while the front will have a high frequency. If both parts of the bike had the same frequency, then every time a wheel deflected to one side it would tend to get into resonance with the other and each sideways movement would promote a similar sideways effort at the other end of the bike – which would be truly unstable.

As the natural frequencies are bound to be very different, this isn't likely to happen. But one wheel could hit a harmonic of the other. (In pure resonance, each left-ward movement of the front wheel would be reinforced by a similar shove from the back wheel, increasing the amplitude of the oscillation very rapidly; if it were a harmonic then every – say – second or fourth movement would be reinforced.) If the force which controls the frequencies goes through a big enough range, there is a good chance that this could happen.

Of course there will be a wide range of frequencies which are 'out of step', that is, one oscillation would oppose the development of the other, and there may be regions where the oscillations are damped out by greater forces at the tyres, etc. But it is enough to explain how steering instability can be so clearly speed-related and to explain how it can develop without a large initial deflec-

* The bike is also stabilized by the wheels (plus the engine shafts) acting as gyroscopes. If a force tries to make a spinning wheel turn left, the wheel will try to lean right; leaning right makes it want to turn right; turning right makes it want to lean left; leaning left makes it want to turn left and so on... During their research into anti-lock braking systems, BMW came to the conclusion that if the front wheel was locked for more than 0.5 seconds then the loss of gyroscopic force would be enough to lose stability and steering control.

tion. Machines with very low steering inertia may have similar problems because the steering can be deflected easily, by a bump for example, and if this happens at a speed/thrust region where one wheel has a sympathetic frequency to the other then what might have been a single twitch of the steering could break out very quickly into a series of increasingly large wobbles.

Where this is a regular problem, it should be possible to alter the characteristics enough by changing the weight distribution (particularly by removing weight from above the pendulum axis or weight at a large distance from the axis) or by altering the castor/trail dimensions, by changing the front or rear ride height. If a machine's steering suddenly goes unstable, the best corrective measure the rider can take is to lightly use the rear brake. This will achieve several things: it will alter the thrust/drag forces, it will reduce the overall speed of the bike and it will slightly increase the front axle loading, all of which alter the mechanism which drives the oscillation, bringing the machine into a more stable region. Accelerating to a higher speed may reduce some of the factors which create instability but the initial thrust force and rearward weight transfer, before the road speed alters significantly, are likely to reduce stability further. Using the front brake could achieve much the same as the rear except that when the steering is fluttering the front tyre is only pointing ahead for a small fraction of the time. For the rest of the time it is generating very large slip angles or may even have lost its tractive grip on the road surface, so trying to put brake torque through it is very likely to make the wheel lock (which may well cure the steering instability, but will not improve the stability of the machine as a whole).

The main types of instability usually take the form of weaving, and there seems to be four distinct groups:

1   Handlebar flutter or wobble.
This usually happens at 35 to 40 mph, starts weak and quickly builds up (within 10 cycles) into a large wobble with a fairly high frequency and large amplitude. I would guess at 3 to 5 Hz and sometimes enough amplitude to use up all the steering lock. It is often highly speed-related and does not have enough energy to build up if the rider keeps his hands on the handlebars. It can be made much more severe if a rigid tail load is fastened to the bike, such as a rear carrier. Some bikes with very light steering do it, especially during hard acceleration over a series of bumps and this is about the only time a steering damper helps. This type of weave can usually be reduced or eliminated by removing weight from the back of the bike, by removing any superfluous mass from the steering, by increasing the trail, by increasing the front axle load.

2   High speed weave
It can happen at any speed but it is usually above 80 mph. It is similar to (1) but usually with lower frequency and stronger so that the rider is unable to prevent it merely by holding the handlebar. It often begins at a clearly

defined speed and sometimes stops at a clearly defined speed (but often the bike will not go fast enough to reach this upper limit, so it is not certain).

It is frequently related to tyres, especially worn rear tyres and the speed at which the weave starts often reduces as the tyre wears down. During some endurance tests on a Honda CB900 at Snetterton five rear tyres were worn out during a 24-hour period; on most of them the Honda would start to weave at about 120 mph, which it reached at the end of the back straight. As the tyre wore down, the weave set in at a lower speed and therefore happened earlier along the straight. When it started to weave at the beginning of the straight, the rider knew that it was time to come in for a tyre change.

One theory is that any castoring wheel will become unstable at some 'critical' speed, which seems to be a likely explanation in this case. It is possible to raise the critical speed by modifying things, ultimately to raise the critical speed above the maximum speed of the bike.

Factors which alter this kind of weave are:

(a) Tyre profile or construction.
(b) Inertia about steering axis.
(c) Frame or swing arm stiffness.
(d) Mass carried high above the wheelbase and rigidly mounted.
(e) Steering castor and trail.
(f) Rear suspension spring/damping.
(g) Steering head bearings adjusted too tightly, perhaps, or some other restriction such as tight or chafing wiring.

3 Suspension-related weave

This problem usually happens with heavy bikes in long, fast curves, when the whole machine seems to rise and fall, swaying from side to side at the same time. A following rider can often see daylight between the wheels. The cure lies in making it lighter and stronger, fitting better tyres and suspension, that is, getting a new bike, which is exactly what the Japanese manufacturers did from about 1978 onwards and since then, standard roadsters have not had this problem.

4 Steering-related weave

A short, sharp sideways movement at the steering head, when banked over, often with no appreciable steering movement. Usually caused by badly-adjusted steering bearings, worn races, cables or wiring chafing on the steering, too much steering damper, tyre or wheel runout or a mismatch between the tyres and the bike.

The first two types of weave happen on the straight, while the second two happen in corners. Other problems in corners are usually due to suspension movement (bouncing) or lack of frame stiffness, but induce some roll and yaw because the bike is banked over. This is usually attributed to poor suspension or a mismatch between suspension and tyres.

Tyres are by far the most important single component. They form part of the transmission, braking, suspension and steering systems and the bike should literally be built around them.

A development programme would take this form:

1 Choose the best tyres for the purpose (Chapter 3).
2 Select wheels, brakes, forks and swing arm to suit (Chapters 4 and 5).
3 Choose the stiffest, lightest frame available, or lighten the existing frame as far as possible (Chapter 8).
4 The hard part: make the unsprung mass as light as possible (Chapters 4, 5 and 6).
5 Choose suspension which will cope with the ratio of sprung to unsprung mass, and the sort of terrain the bike will be used on (Chapter 5).
6 Work out the weight distribution and steering geometry, bearing in mind the power output of the engine (Chapter 2).
7 Improve the aerodynamics, provide air ducts for cooling, engine intake, etc. Will probably need a series of tests later (Chapter 7).
8 Bodywork and riding position (Chapters 8 and 9).
9 Set up tests to evaluate the most important aspects, for example traction for a motocrosser, cornering traction for a road racer, etc. (Chapter 11).
10 Fine tune the handling responses, etc. – which may need changing for different circuits (Chapters 2, 5, 8 and 11).

The object is to maximize grip and then maintain it under all conditions, to improve performance by making the bike lighter and more slippery, to make it as easy as possible for the rider to handle and finally to prevent it from destroying itself or the rider.

Chapter 2

Rolling chassis

Before any modifications are made to a bike, it helps to have a large, fairly accurate drawing of the machine. Then, when something is altered, the whole chain of consequences can be seen in advance. Most changes, like wheel size or spring length, have several other effects, for example on the trail, ground clearance, chainline clearance, braking ability and so on.

Measuring directly from the bike is not easy so, unless you are very thorough or are able to make up a suitable jig, such measurements are likely to be inconsistent and should only be treated as approximations.

You can, however, make accurate measurements of parts which are of special interest simply by marking suitable reference points on the machine itself. These points must be accessible and capable of being measured easily using a millimetre scale, a made-up gauge, a go/no-go gauge, etc. The measurements in themselves will have no meaning, except that they will tell you how something has changed. A typical example is front and rear ride height. Instead of measuring vertically up to the sprung end of the suspension, it will be much easier to locate two accessible parts, one on the sprung and one on the unsprung part of the suspension, so that it is convenient to measure the gap between them. Sticking masking tape to the frame and drawing a cross on it will pin-point the spot.

A drawing of the bike will show suitable locations which are not only easy to measure but also show the maximum change (for example, the end of the swing arm moves further than any other point on the arm, for any given amount of suspension travel). The consequent figures for ride height or ground clearance have no meaning on their own but they will reflect changes very accurately. Also, it is one thing to be able to take measurements in your own workshop and a totally different matter to try to do the same thing in a racetrack paddock where the ground is uneven, you are stepping over puddles and everybody is in a hurry.

Having reference points marked on the bike, or noted in advance, will be a big help when the testing gets totally confused and you want to set it all back to where you started.

Figures 2.1 and 2.2 give some idea of the type of reference measurements which can be made. Front and rear ride heights need to follow the suspension as closely as possible and must be made with a known load on the bike. The position of the forks in the yokes can be measured with a vernier caliper or depth gauge – sometimes there is a reference line at the top of the fork leg. When the bike is set up in the workshop the forks will be drained of oil, and the required amount will then be measured into each leg. This is a good time to make up a dip-stick to measure the oil level with the fork in a repeatable position, and to see how the level varies when different volumes of oil are used.

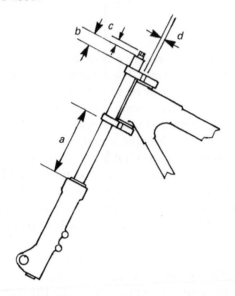

Figure 2.1 Reference measurements on the steering and front suspension. *a* Ride height; *b* Height of forks in yoke; *c* Pre-load adjuster (or length of spacer on top of spring); *d* Adjustable yoke spacer or head bearings carried in eccentric cups. It is also worth measuring the height of the oil from the top of the fork leg with the spring removed and the fork (and damper rod) fully compressed

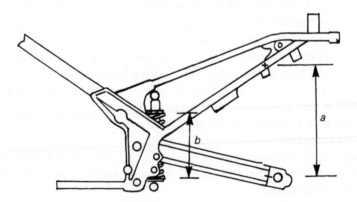

Figure 2.2 Reference measurements at the rear suspension. *a* Ride height (between two datum marks); *b* Fitted length of spring

During testing it is much easier to measure the oil level than to rely on volume measurements, mainly because it is difficult to drain the oil completely without stripping the forks down.

18

Figure 2.3 Adjustable fork yokes made by Spondon Engineering. The thickness of the spacer (clamped between the two halves of each yoke) determines the fork offset and the trail

In the same way, the wheels should be aligned precisely and then the marks on the swing arm calibrated, or the screw adjusters ground down so that they are of equal length when the wheels are in line and can be checked using a pair of calipers. It is easier to align the wheels using builders' line or a straight edge before the tyres are fitted, so that the measurements are taken from the hard, straight edges of the wheel rims, even though the rear one will probably be larger than the front and there will have to be some means of spacing the front accordingly.

Castor and, more importantly, trail are virtually impossible to measure with the accuracy which is needed unless the frame can be held in a jig. Consequently it is important to be able to measure other factors which can alter the trail so that you will know whether it has been increased or reduced. Another example of when a scale drawing is useful.

Trail is probably the most significant dimension, once the bike is constructed and ready for testing, along with weight distribution. Other items, like suspension changes are only important in the way that they may alter this basic geometry – but there will have to be changes to get a better ride, to improve ground clearance and so on. Each time the overall effects on the steering geometry and the way it influences handling need to be considered.

Weight distribution is significant: it affects the ability to brake, accelerate and to corner and, on some competition bikes, it is changed to suit individual circuits. It (or the position of the bike's centre of gravity) crops up all the time and is quite easy to measure using ordinary bathroom scales.

1 Use two scales or raise the non-weighed wheel to the same height as the other, using blocks of wood, etc. The bike must be on level ground, or the wheels packed up until it is level. To check that the scales are level, put the bike on them, get front and rear weights, then turn the bike round. If the scales are not level the second front/rear figures will be different.
2 Calibrate the scales using dead weights, or get yourself weighed accurately and use the scale adjusters to read your true weight. At least make sure they both read the same when they carry the same weight.
3 Weigh the front and rear wheels on level ground. The bike needs to carry its usual load of fuel, oil and the rider in his normal position.
4 Repeat these measurements until they are consistent.
5 Raise the front, using wooden blocks (weigh the blocks and subtract this from the front axle weight). The higher the wheel is raised, the bigger the front/rear difference, so the result is less affected by weighing errors. However, a greater height will cause more suspension movement, all liquid levels to swill backwards and more difficulty for the rider. The optimum amount is 6–10 inches. Use horizontal straps to balance the bike, as long as they don't interfere with the weight measurement. (We made an adjustable stand, in the shape of a letter A, with tapered pins at the apex which push into the hollow rear wheel spindle. The horizontal bar is a large screw thread that is unwound until the legs balance the bike but all the weight is carried on the tyre).
6 Measure the wheelbase and the rolling radius of the rear tyre.

If the bike's wheelbas is w, let the centre of gravity be a distance x horizontally from the front wheel spindle, and a distance y vertically from the ground. The weights carried on the front and rear axles will be denoted by F_1, F_2 and R_1, R_2, the suffix 1 being for the level measurement and 2 for the measurement with the front wheel higher than the rear.

From the horizontal measurement, we can get the weight distribution, F_1/R_1 or as a percentage:

Front = $100\ F_1/(F_1 + R_1)$
Rear = $100\ R_1/(F_1 + R_1)$

We can also find the value of x:

$$x = R_1 w/(F_1 + R_1)$$

If the height of the front wheel above the back wheel is h, and the rolling radius of the wheels is r_f and r_r then the height of the centre of gravity is:

$y = r_R + [w - x) - D/\cos \theta] \times \tan (90 - \theta)$
where $D = w \times \cos \theta \times [1 - R_2/F_1 + R_1)]$

20

where $\sin \theta \approx h/w$ (this approximation loses accuracy if the radius of one wheel is significantly larger than the other)

From these equations it is possible to get values for the height and position of the centre of gravity and, in the process the rolling radius of the tyres and the total weight of the bike will be found. These figures can be used to calculate the bike's potential for acceleration, for braking and whether anti-dive suspension would be beneficial or not (see Appendix and Chapter 6). It is also possible to see the effect of moving the centre of gravity – up, down, backwards, forwards – until the best traction is found. These considerations will affect the choice of engine mountings, length of swing arm, etc. Once the best centre of gravity position has been decided, it is a matter of getting as

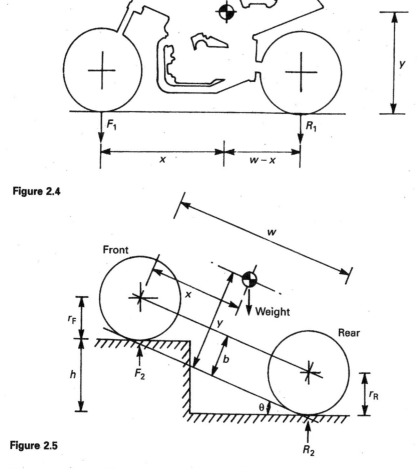

Figure 2.4

Figure 2.5

Figures 2.4 and 2.5 The centre of gravity can be measured by weighing the bike on level ground and with one wheel raised. This is the notation used in the text

close as possible to this and still have adequate ground clearance, optimum wheelbase for handling, and so on.

The further back the centre of gravity is, the better the rear wheel traction because there is more weight on it. This also increases the tendency to wheelie under acceleration, as does a high centre of gravity because the overturning couple acting on the rear tyre's contact patch is greater. So, more rearward weight transfer gives more traction and increases the likelihood of a wheelie. Moving weight forward and lower down reduces both of these possibilities and increases the chance of wheelspin.

In the Appendix, the program RL calculates acceleration and predicts wheelies or wheelspin as a limiting condition. The calculation is straightforward but tedious to work out for all engine speed/power combinations. Using the program it is possible to try various positions for the centre of gravity, alter the wheelbase, change the tyre friction, rolling radius and the aerodynamic properties of the bike. The results are shown as acceleration from zero to maximum speed.

Optimum centre of gravity

The position of the centre of gravity determines how much weight transfer there is during braking and acceleration. It therefore controls the amount of traction available at each wheel.

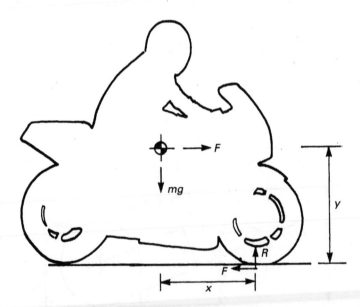

Figure 2.6 Overturn forces. Acceleration and braking forces (*F*) are generated at ground level while the inertia of the bike is equal in size, opposite in direction and acts through the centre of gravity. The height of the centre of gravity produces a couple *Fy* which tries to overturn the bike and which shifts more load on to the rear axle during acceleration and to the front axle during braking, so the vertical reaction at the wheel (*R*) tends towards *mg*, the weight of the whole machine

If we assume that the ideal condition is to have the rear wheel start to spin just as the front wheel begins to lift during acceleration (and, during braking, that the front wheel will lock just as the rear wheel starts to lift) then the geometry of the bike can be arranged to give these conditions. The reasoning is: (a) acceleration – this is maximum traction, no more power can be transmitted; (b) braking – the maximum level can be reached by using front and rear brakes, but as the rear wheel load is reduced the rear brake becomes more difficult to control without locking. The solutions are to use the front only (with geometry to maximize front wheel traction), or to develop geometry which does not permit so much weight transfer (long wheelbase, low CG), to keep the rear brake easy to use, or to develop antilock brakes.

If the mass of the bike is m, its centre of gravity is located distance x behind the front wheel, y above the ground and the wheelbase is w (see Figure 2.6), then the forces acting on the bike can be calculated. The reactions between the ground and the tyres are R (front) and R_1 (rear) and the coefficient of friction between tyre and road is μ.

Acceleration

The maximum force that can be transmitted at the rear tyre is $R_1\mu$, beyond which the tyre will spin. If the front wheel is also lifting at this point, then $R_1 = mg$, that is, all the weight is now carried on the back wheel.

The couple trying to overturn the bike is $R_1\mu y = mg\mu y$ and the effect of the bike's weight, opposing this is $mg(w - x)$ and the two are just balanced, so:

$$mg(w - x) = mg\mu y$$

$$w - x = \mu y \tag{2.1}$$

Braking

The maximum force transmitted by the front tyre is $R\mu$ which becomes $mg\mu$ if the rear wheel is beginning to lift. The overturning couple is $mg\mu y$ as before (but in the opposite direction) and the stabilizing effect of the weight is mgx, so:

$$mgx = mg\mu y$$

$$x = \mu y \tag{2.2}$$

From equations 2.1 and 2.2:

$$y = w/2\mu \text{ and } x = w/2$$

Therefore the centre of gravity should be midway along the wheelbase and its height depends on the wheelbase and the friction available at the tyres. If $\mu = 1$ (sports compound, dry track) then $y = w/2$. The height of the centre of gravity is also half the wheelbase. If $\mu = 0.5$ (road tyres, wet road) than, ideally, $y = w$.

Note: There is a second-order torque caused by the acceleration of the wheels. The reaction to this, depending on the inertia of the wheel and its acceleration, will add to the overturning couple in both cases but is not allowed for in these equations.

This is not totally practical but, when friction levels are low, the rider can improve traction by (a) raising the centre of gravity (sitting up on a road racer, standing up on a dirt bike) and shifting back for acceleration or forwards for braking (changing the value of x). The limiting condition is likely to be wheelspin or a locked front wheel. During braking the front will lock long before the back is likely to lift; there will be weight on the back wheel so more braking effort can be found by using the back brake as well as the front in slippery conditions. (Note that this will cause more weight transfer, which will permit more front brake force to be transmitted.)

Finally, these proportions x and y take no account of suspension movement caused by the weight transfer. Therefore the dimensions x and y give the position of centre of gravity *during* acceleration or braking. Braking makes the front suspension compress, and the centre of gravity will be lowered (also the wheelbase will be shortened if telescopic forks are used), so the static values for y and w will be slightly greater than the values found from the above equations. Similarly, the rear suspension will compress during acceleration.

Rather than try to calculate the suspension travel, it is easier to measure it (a smear of grease or a cable tie on the fork stanchion or damper rod will show the extent of travel) and then use a drawing to find the movement of the centre of gravity which this causes. To get the centre of gravity at the optimum height, there may be a need to make the bike taller or lower, to change the wheelbase, or to incorporate some measure of anti-dive or anti-squat suspension (see Chapters 5 and 6).

These considerations are for straight line acceleration and braking. During cornering there is a need for equal grip at both wheels, so a centre of gravity near the centre of the wheelbase is desirable. The height of the centre of gravity is open to compromise. The higher it is, the more weight transfer there will be and weight transfer is necessary in proportion to the amount of power/braking which is being transmitted, so that the relevant tyre is given more traction. A higher centre of gravity also needs a slightly smaller angle of lean – something which is increased by the use of wider tyres.

The benefit of a low centre of gravity is that the mass of the bike will have less inertia about its roll axis (which will be at, or just above, ground level). Less inertia means that the bike will roll more readily and will be more responsive to steering input. Also, once rolling, it will be easier to stop or to reverse the movement and this also applies if the bike slides, developing sideways inertia.

During cornering the tyres have to cope with cornering forces as well as power transmission, so it is likely that the limiting condition will be a slide/wheelspin rather than an overturn or wheelie. Experimentally, if the wheel tends to lock or to spin too easily then the centre of gravity should be raised and/or moved towards that particular wheel. If the bike raises one wheel as a limiting condition then the centre of gravity should be lowered and/or moved towards the wheel which lifts. (Bikes often are able to wheelie during cornering; the equivalent during braking is to lift the back wheel but this would normally appear as the back breaking away when the front brake is used in a corner.)

To set up the rolling chassis there are a few adjustments which need to be made and several other changes which can be used to alter the handling characteristics and which should be measured before testing the bike. These are:

1 Wheel bearings
Use a wheel bearing grease and make sure that the bearings are accurately spaced and that the wheel spindles are spaced so that when they are clamped there is no tension in the forks. Bending in the fork legs will interfere with suspension movement and can misalign the calipers, causing brake drag. See Chapter 4 for details on balance, runout, etc.

2 Wheel alignment
Space the sprockets so that the chain line is straight, then set the wheel alignment accurately using the wheel rims with no tyres fitted. The rear wheel should be on the bike's centreline and the front wheel should align with it exactly. It will probably be necessary to make up spacers to allow for the different rim widths, so that a straight edge or builder's line can be used, as high up the rim as possible so that the two contact points are as wide apart as they can be. It may be necessary to grind spacers to get the exact alignment – check that this does not interfere with brake alignment and that there is clearance at the chain and torque arm when the tyre is fitted.

Where a very wide rear tyre is used, it may be necessary to move the sprockets over, using an offset gearbox sprocket, a longer gearbox shaft or moving the engine unit. As the use of a wide tyre implies a lot of power, this may also mean that the gearbox sprocket/shaft will have to be supported on an outrigger bearing. Do not be tempted to move the rear tyre away from the centreline; it is better to move the engine to one side.

Once the wheels are accurately aligned, calibrate the scale on the chain adjusters, make a new scale or make the adjusters equal in length so that they can be measured whenever the wheel is disturbed in the future.

The changes in chain tension should be checked over the full suspension travel before the spring is fitted. A tensioner or protective strips may be needed – tough plastic strips are fitted to many machines as standard and it should not be difficult to find something that can be adapted. Tension the chain with the suspension in its tightest position; the top run should not quite go tight. Then see how much slack this gives when the bike is in its normal position so that in the future the chain can be tensioned to this level.

3 Suspension bearings

Clean and grease frequently, use the grease to fill any holes or spaces to keep out mud and water. The bearings are normally set with zero end float and no preload; it may be necessary to make up shims or spacers to achieve this. Some taper roller bearings are fitted with a pre-load, so check with the manufacturer's manual first.

Figure 2.7 Outrigger bearing to support the gearbox output shaft, fitted to a special built by Steve Burns

4 Steering bearings

Adjustment is critical. Remove any restriction such as wiring or control cables and consider permanently re-routing any heavy cables which have to move with the steering.

Jack the front end up until the suspension is completely extended; loosen the bottom yokes and any locknut on the steering stem adjuster. Tighten the adjuster until the steering is just free to move under its own weight, with no detectable play. If there is any pitting or stickiness in the bearings, fit new

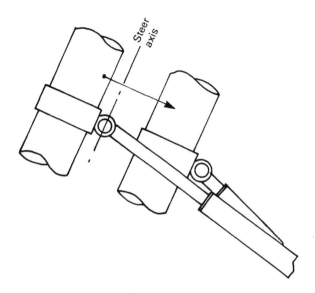

Figure 2.8 A steering damper needs to be clamped so that the moving end is at right angles to the steering axis, otherwise the damper body will twist when the steering is moved. Alternatively, the ends should be carried in spherical bearings

ones. Be careful not to move the adjuster nut when the locknut is tightened. When taper roller bearings are first fitted, the steering stem is usually over-tightened to pull the bearings on to their seats, and then slackened before the adjustment is made.

Ball races are less sensitive to adjustment and are arguably better at taking axial loads. Taper rollers need very careful adjustment and can cope with radial loads better (for example brake forces). A combination of bottom ball-race and top taper roller is a possibility. If the running conditions have lots of severe bumps, wheelies or jumps then the races' may soon show damage – usually pitting – and will have to be replaced.

5 Steering damper

Its only use is to stop steering flutter when the front wheel goes light over bumps at high speed or during acceleration. It will not cure weave or other stability problems and may make them worse. It should either be fitted so that the rod is at 90° to the fork leg (to the steering axis, to be precise) or so that the body and the rod end are mounted on spherical joints, otherwise the steering movement will try to bend the rod.

6 Steering inertia

Steering response and stability are both proportional to the mass carried on the steering and its distance from the steering axis. These parts should be

27

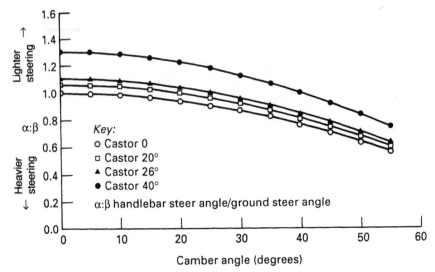

Figure 2.9 The ratio between handlebar steer angle and ground steer angle (α:β) for a range of camber (lean) angles and castor angles

removed, relocated or made lighter:

- Handlebar length and material – make shorter, lighter (note the use of weights to damp out vibration – see Figure 8.11).
- Mirrors, indicators – move to fairing.
- Instruments, lamps – carry on frame-mounted bracket or in fairing.
- Levers, master cylinders – use lighter components, consider a remote reservoir mounted close to steering axis (possibly shared by hydraulic clutch and brake).
- Mudguard, fork brace – make as light as possible.
- Brake calipers – use light material, mount as close to steering axis as possible (that is behind fork leg, usually).
- Wheels, tyres and discs – as light as possible, given the sizes that are deemed necessary.

7 Front ride height

This can be altered by fitting a different length spring, a different spring rate or a different pre-load. It can also be altered by moving the forks through the yokes. The main effect is to alter the castor and trail (lowering the front gives steeper castor, less trail and will tend to make the steering lighter, faster, less stable). It also changes the ground clearance and the frontal area. The fork/yoke position should be measured as should the distance between the unsprung part of the fork and some convenient part of the sprung mass (usually the bottom yoke). It may be necessary to find out if this distance alters when the bike is being ridden (owing, for example, to aerodynamic lift or to the suspension pumping down over a series of bumps).

8 Rear ride height

Changing the spring rate, length or pre-load will alter this, as it did for the front. It is also possible to fit longer spring units or to fit an extension to the body of the damper, or to change parts in the suspension linkage to alter the ride height. Raising the ride height makes the castor steeper and gives less trail (faster steering); it also increases ground clearance. It is usually most convenient to measure it from the top of the swing arm, near the wheel spindle, vertically up to some convenient part of the bodywork.

9 Wheel size

A wheel's inertia is proportional to its weight and to the square of its diameter, so big changes can be made by fitting smaller wheels. The diameter of the wheel and tyre also affect the ground clearance, gearing (rear), speedo drive gearing, probably the castor and trail (a scale drawing will show how much), the braking effort (depending on the disc's diameter, see Chapter 6), so there can be some fairly fundamental changes. There may also be clearance problems at the chainline, brake torque arm, mudguard, etc., especially if tyres grow at high speed (cross-ply tyres change most, bias-belt and radials change less).

10 Trail

On some machines this is directly adjustable. Basically, more trail means that the tyre's contact patch is deflected further for a given steer angle; the steering effort is therefore greater, the self-aligning torque (which will pull the wheel back into line if you release the handlebar) will be greater. The steering is heavier and more stable if trail is increased.

11 Castor

This is also directly adjustable on very few designs. It translates handlebar steer angle into wheel steer angle: very steep castor translates the handlebar movement directly; very shallow castor does not turn the wheel so far but it does make it lean more when it is turned (an angle of zero to the horizontal would give no turn at all but the tyre would lean by whatever angle was steered at the handlebar). It is measured either from the vertical or the horizontal. In the common, 60° (from the horizontal) region, there is a blend of turn and lean which combines with the amount of trail to produce the steering force which we feel.

The steer angle at the wheel is related to the steer angle at the handlebar by the castor and the camber of the steering axis. If α is the steer angle at the handlebar, β is the steer angle at the wheel, θ is the castor measured from the vertical and ϕ is the camber measured from the vertical then:

$$\beta = \alpha \cos \theta / \cos \phi$$

Steeper castor makes the steering more direct but heavier.

12 Swing arm length

Changing this length alters the wheelbase. A longer swing arm makes a larger turning circle for a given steering deflection. As the bike is less sensitive to steering input, it becomes more stable and, by the same token, less manoeuvrable.

The change in wheelbase also alters the weight distribution considerably. A longer swing arm means less weight is carried on the rear wheel which is therefore more likely to spin. If the weight and centre of gravity is already known, this can be calculated fairly easily (see Chapter 5).

Operating the suspension through a longer lever effectively makes the wheel rate softer, so a longer swing arm also gives softer suspension. If the swing arm is not horizontal then this, plus the effect on the suspension leverage, will also change the rear ride height and the tendency to squat under acceleration.

13 Engine plates

Moving the engine back and forth is a useful way of changing weight distribution. Competition bikes sometimes have various mountings so that they can get more or less traction depending on the circuit.

14 Riding position

As the rider makes up almost 50 per cent of the total weight of a competition machine, it can be easier – and more effective – to move him around to alter the weight distribution. It may mean moving the seat, handlebars and footrests to preserve a comfortable position but, especially during the development period when control layouts can be less than perfect, it is a valid method for finding more traction.

All of these components should be set up before the bike is fully built, the settings marked or noted and any side effects taken into account. To tune the handling, trail (which is tangled with castor) and weight distribution are the important things. Everything else is secondary – or has some other function, such as suspension – but it can (deliberately or unwittingly) alter the main steering geometry and therefore affect the handling.

Steering reaction

If the steer angle on the ground is β then the steer angle at the handlebar, α, is given by:

$$\alpha = \beta \cos \phi / \cos \theta$$

Where ϕ is the camber angle and θ is the castor, both measured from the vertical.

The rider has to provide a force F on the handlebar, at a distance x from the steering axis, so the steer torque is Fx.

This balances the reaction between the tyre and the road which is made up of several components. The centre of pressure of the contact patch is displaced a distance y, so that $y = t \sin \beta$ (where t is the trail) caused by the steer angle. It is also displaced, in the opposite direction, by the effect of the bike leaning and the contact patch moving around the shoulder of the tyre, see Figures 3.5 and 3.6. The total displacement of the centre of the contact patch is

$$y = t \sin \beta - r \tan \phi$$

Where r is the section radius of the tyre at that lean angle.

In terms of handlebar steer angle:

$$y = t \sin [\alpha \cos \theta / \cos \phi] - r \tan \phi$$

The various forces acting through the contact patch set up torque about the steer axis, namely:

Axle load, mg, creates a torque $mgt \sin \phi \cos \theta$, turning the steering in; centripetal acceleration, a, caused by force ma, creates a torque $mat \cos \phi \cos \theta$, turning out; rolling drag, brake force, D, creates a torque $Dy \cos \theta$, which may turn the steering in or out, depending on which way the contact patch is displaced.

In a steady state:

$$Fx = mgt \sin \phi \cos \theta \pm Dy \cos \theta - mat \cos \phi \cos \theta$$

From this it may be seen how the force necessary at the handlebar depends on the bike's geometry and cornering attitude. According to the relative size of each component, Fx may be positive, zero or negative. It may even change from negative to positive if one of the terms alters sufficiently (typically when the brake is used, creating a torque which makes the steering try to straighten up).

Two-wheel steering

For competition use it is usually an advantage to make the steering as light and fast-responding as possible. The reason is to cut down on the time it takes between the rider initiating the turn and the bike adopting a settled, cornering attitude. However this can make the steering unpleasantly sensitive – 'twitchy' – and get very close to making it unstable. A possible solution to this is to steer the rear wheel as well as the front. The rear wheel can be turned the same way as the front ('same phase') or turned the opposite way ('opposite phase'). Same phase steering tends to dull the response and should increase stability: opposite phase steering should have the opposite effect – a sharp response and less stability. It could be used in one of two ways:

1   To allow quite conservative (stable) front steering geometry and achieve the rapid response by using opposite phase steering.
2   To use radical front geometry to get light and fast steering and damp out the unwanted side effects by having same phase rear wheel steering.

Engineers at Yamaha, Atsushi Matsuda, Toshiyuki Satoh and Akira Hasegawa built some experimental prototypes to investigate these possibilities. The rear wheel was carried on a stub axle with a CV joint built into it and was controlled by a cable-operated linkage connected to the front forks via adjustable bell cranks, so that the steer ratio could be altered (i.e. the amount the rear wheel turned in relation to the front wheel).

This was varied from −0.1 to +0.4 (i.e. 1/10 the amount of front steer in the opposite phase through to 4/10 of front steer angle in the same phase). They found that the steering was altered as predicted – opposite phase making it sharper, same phase making it heavier, slower to respond and more stable.

At levels greater than +0.2 steer ratio the machine gave a 'sensation of some disharmony' – the handlebar resisted steering and the bike wanted to sit up in turns – yet below this value their test riders said it gave a good response. When the steering was adjusted to opposite phase, even though it was a very small steer ratio, there were clear benefits in low speed manoeuvrability.

In off-road testing (the prototypes were based on small dirt bikes) the opposite phase made the steering too sharp, making the bike difficult to control and getting worse as the speed increased. The test riders preferred a small (+0.1) amount of same phase steering, which made the steering a little heavier but the machine was easy to handle and gave an impression of improved stability (motocross bikes, although very light, usually have very stable steering geometry, that is a relatively long wheelbase and trail combined with castor which is not as steep as on road or racing machines).

It seems likely that an intelligent control system would be needed so that the appropriate phase could be selected, either to sharpen the steering response for rapid manoeuvring or to dull it for maximum stability.

Chapter 3

Tyres

The tyres are the most significant part of the whole machine. The rest of the bike should be designed around whatever is the best set of tyres for the job. The choice of tyres will dictate rim width, brake size, forks, yokes, swing arms, chainline (and therefore possibly engine mounts) and will even affect steering geometry. They are fundamental to the whole thing and can make changes which reflect this.

Choosing tyres is never easy. The only sure way is to test them on the machine and in the environment in which they will be used (see Chapter 11). Other people's tests and magazine tests are a useful guide, but beware of taking too much notice of tyres supplied to sponsored riders as they are often not the same as the ones on sale, even if they have the same pattern and markings. And, if the manufacturer has made a wise choice of rider, there is a fair chance that he will go well on practically anything.

A list of nominal sizes is given by each manufacturer – this is based on the standards to which tyres are built but real tyres often vary from the dimensions given and will vary again depending on the rim fitting, load and pressure (and the operating speed in some cases). For competition use, the range of tyre types can be staggering and the only people who can usually make any sense out of the array which is available at any one time are the tyre companies' competition reps and their specialist suppliers who run a racing service. They should at least be able to bring the choice down to a number small enough for the rest of us to grapple with.

Size obviously controls the amount of rubber in contact with the ground, especially when the machine banks over. The size and shape of the contact patch are important, but there is more to it than that. The tyre is circular when viewed from the side. It is also, approximately, circular in an end-on section, yet it has to deform to being flat where it meets the floor and has to be capable of doing this when the wheel is banked over to the side. This continual flexing makes the tyre get hot. Also, although it has to be flexible, the tread has to be strongly braced where it is in contact with the ground. Any wrinkling, buckling or arching of the contact area would give a loss of grip. As shown in Figure 3.1, the tyre compresses from a free radius R_{free} to the minimum rolling radius R_{min} in the time it takes the tyre to roll through half of the contact patch. The tread is then expanded back to its normal shape in a similar time. When the wheel is banked over, these compressive/expansive cycles can be more extreme. The strain at the tread surface can often be seen in the way the individual tread blocks wear, tapering on the leading edge and rolling little balls of rubber off the trailing edge.

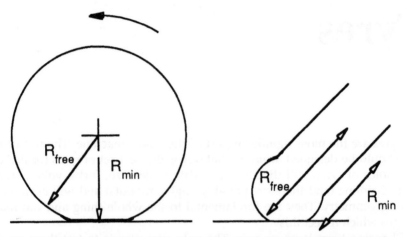

Figure 3.1 A rolling tyre deforms to suit the road surface at its contact patch, from its free radius down to a minimum radius at the centre of the contact area. The surface speed of the tread must reduce by the same proportion so the tread is subjected to a compressive force up to the centre of the contact patch and then a tensile force as it is accelerated away

Data recorders using infra-red sensors to monitor the surface temperature of the tyre show large increases during cornering followed by decreases on the straights. One implication of this is that a varying amount of slip between the tyre and the road surface takes place throughout the contact patch: under conditions where the forces are increased by braking, accelerating or cornering, the forces may not be symmetrical about the centre point of the contact patch and the centre of pressure may move away from the geometrical centre of the area.

Finally, the side walls need to be stiff in the horizontal plane so that they do not let the wheel twist or lean to the side, relative to the tyre.

The tyres can, and do, flex vertically up and down which, with the compression at the tread, acts in series with the suspension. Because the tyre acts like a spring and takes the same loads as the suspension, the two have to be adjusted together – a change to a different type of tyre is quite likely to need different suspension settings. (If the suspension is not optimized for each type of tyre the ride and handling will not be at its best, which does not make comparative tyre testing any easier.)

The sidewalls and shoulders of the tyre have some stiffness even if the tyre is not inflated. If the tyre is then inflated to a pressure p, its contact area will be given by:

$$pA + s = W$$

or:

$$A = (W - s)/p$$

34

where A is the contact area, s is the sidewall and crown stiffness and W is the weight on the axle.

Reducing the pressure will therefore tend to increase the contact area, A. The tyre depends on the pressure for its stiffness, so lowering the pressure too far will allow deformation at the contact patch and in the sidewalls. It follows that there will be an optimum pressure for which the contact area and grip will reach a maximum. At higher pressures the contact area will get smaller, at lower pressures the tyre will flex too much. This optimum pressure depends on the axle load, W. Fitting a larger tyre, which makes A larger, will require a smaller pressure. Physically, the largest acceptable tyre is one whose contact area is so great that the required pressure is at the lowest possible setting before the tyre gets too flexible.

Raising the pressure reduces the contact area and therefore reduces the total amount of deformation for each revolution of the wheel. Higher pressures therefore encourage the tyre to run cooler, and vice versa.

Very low pressures in off-road tyres let the tyre deform so that it can follow the contours of the ground more easily and the cyclic tread deformation reduces clogging by mud.

The size and shape of the contact patch depend on the tyre's stiffness (that is, its internal strength plus the effect of the air pressure inside it), and its original shape (that is, its profile). The profile mainly governs the way the contact patch changes as the bike rolls into a corner. The profile will change with different pressures, with different rim widths and will change as the tyre tread wears down. So for optimum grip and handling characteristics, the tyre needs to be fitted to the correct rim size as well as being run at the optimum pressure (and not badly worn).

Typically, crossply tyres in /90 or /100 sections up to 140 size may be fitted to three or sometimes four rim widths. The wider the section and the lower the aspect ratio, the more critical it becomes, until the low-section, low sidewall radial ply tyres (for example 170/60VR 17 or wider/lower) should only be fitted to one rim size.

Tyre size 130/80VB 18 explained

130 is the nominal section width in millimetres.

/80 is the aspect ratio (height of the section as a percentage, i.e. a height of 104 mm above the rim seat in this case. No number means 100%).

V, the first letter, is a speed rating. V means up to 130 mph.

B, the second letter, refers to the construction. B means biased belt, R means radial ply and nothing or '–' means cross ply.

18 is the wheel rim diameter in inches.

Racing tyres have different size codes, depending on the manufacturer. For example, Dunlop might describe the section as 325/425, which is the (nominal) height/width in inches. Michelin use a code such as 15/61, which means width (in centimetres)/diameter (of the tyre, also in centimetres). Both are nominal.

The shape of the contact patch and the way in which the tyre deforms are critical to its performance. It gets its strength from 'plies' – sheets woven from nylon, rayon, kevlar or steel. The woven material usually has warp and weft threads at approximately 90° running under and over one another. When this woven material was first used in tyres it gave the required multi-directional flexibility but the double thickness where the warp crossed the weft thread caused chafing, fraying and unpredictable failure. A weftless cord material was subsequently developed, with all parallel warp threads held together by thin 'pick' threads every half an inch or so across the material, to allow it to be handled before assembly into the tyre drum. The plies are anchored around steel cables which run around the bead at each edge of the tyre. The material, the number of plies and the angle at which the threads run all contribute to the strength of the carcass and these plies are called carcass or casing plies. When they are set at an angle to one another, so the lines of the threads cross at quite a steep angle, the construction is called cross-ply or diagonal ply.

The material, the angles of the plies and the number of plies can be changed to alter the strength and the deflection of the tyre under load. In cross ply tyres, four plies are typically used, sometimes with a wedge-section filler to stiffen the sidewall. It makes a supple and cheap, although rather heavy, tyre. Flexing of the necessarily thick material makes it run relatively hot and the deflection of the tread at the contact patch causes deformation in the form of outward bulges fore and aft of the contact patch. This localized deflection makes it harder to control the shape of the contact area as well as making the tyre run hotter. A side-effect of this construction is that the tyre grows, diametrically, under the effect of centrifugal force at high speed. It also contracts across its width. The effect raises the gearing, changes the profile and often wears away parts of the mud guard too.

The first attempts to increase the strength of bike tyres began when kevlar plies were used and when a strong 'belt' was run around the complete tyre, underneath the tread, either at zero degrees or at a shallow angle to the line of the wheel (the bias belted construction). This belt was intended to prevent the kind of deflection that ordinary cross plies suffered, which would give better control over the tyre and allow it to run cooler. It could then be made with a softer compound which would give more grip.

As this development progressed, it became possible to make tyres with fewer elements, the lighter construction gave cooler running and permitted still softer compounds. At this stage, very high angle cross ply construction was being used, only one step away from plies which did not cross at all, but which ran parallel, at 90° to the bead – the radial ply construction.

This has several advantages. Each ply takes the shortest route from one anchoring bead to the other; thus each individual thread is stronger than an equivalent piece of material in a cross ply construction and, for the same weight, there will be less deflection. A circumferential, zero-degree belt gives strength under the tread area and reduces the amount of 'fling' at high speeds. The construction also makes the tyre deform fairly evenly over the whole circumference, instead of the localized bulges near the contact patch which are characteristic of the cross ply. This gives greater control of the tyre shape and also permits cooler running.

A shape like a tyre, curved in two planes, deflecting considerably once per revolution, is a very complicated model to imagine, let alone to do stress calculations for. Early tyre development tended to be empirical and experimental – it was easier to make the tyre and test it rather than try to predict its performance. Computer-assisted design has helped considerably; the designer can now see the effects of changing ply material or angles; he can experiment with the shape on the screen until it is doing what he wants it to, then he can go and get the tyre built. It is a process which must get more accurate, as the results of real tests can be used to correct the predictions of the computer.

Avon used this technique to improve its cross ply design. When a beam is bent, the convex side is in tension, the concave side in compression; somewhere in the middle there is a neutral axis which is neither in tension nor compression. The same applies to the structure of a tyre and if the reinforcing plies could be placed on the neutral axis, they would experience smaller stress fluctuations and therefore run cooler, etc.

But it was the radial ply which held the most potential and *not* for the obvious reason that this design had transformed car tyres, giving better grip and double the mileage of cross plies.

Development for car tyres was based on a totally different need. Here the radial ply tyre was successful because it could brace the tread area flat by using a belt and allow flexible sidewalls so that cornering thrust did not have to be generated by using severe slip angles. In the motorcycle tyre, the sidewall cannot be flexible. To quote François Decima, head of Michelin's Research and Development Centre at Ladoux, 'The sidewall has no function other than to connect the tread to the rim and to bear the name Michelin.'

The easiest way to make the sidewall stiff is to make it short. This also shortens the bead-to-bead distance and improves the whole structure of the radial ply tyre.

Perhaps it was fitting that Michelin, who had developed the steel-braced radial for cars, should also pioneer the radial ply tyre for bikes. Early attempts were made difficult because existing bikes had been designed for high aspect ratio tyres. Fitting the low sections, at which radials work best, reduced ground clearance, changed the steering geometry and even lowered the gearing. Clearly the bikes and tyres had to be developed together.

Michelin concentrated on racing tyres, with one or two ventures into OEM for new roadsters. Other manufacturers (Bridgestone, Dunlop, Pirelli and Metzeler) developed radials, or very high angle cross ply tyres, in the high sections which would fit existing roadsters. These tyres were too compromised; to stiffen the sidewalls, the tyres had to be made heavy. Dunlop, for example, used a thick rubber filler between the bead and the tyre shoulder. Pirelli had two carcass plies, a reinforcing strip on the tyre's shoulder, turn-ups around the bead, two biased belts and a zero degree belt in its early MP7. The results were tyres which had almost as many elements as cross plies and, although there were small improvements, they could not show the full potential of the radial design. A comparison between Michelin's M48, V-rated cross-ply and their 160/60VR M59X radial, shows that the M48 has three casing plies and two crown plies while the M59X has just one casing ply and one zero-degree belt.

The potential could be used when, following racing practice, road bikes were fitted with wide rims and designed for use with low section tyres. Figures 3.2 and 3.3 show the differences. The deflection owing to load is greater on a 160/60VR18 than it is on the taller 160/80VR16, yet nearly all of this deflection is in the shoulder and the crown of the tyre – there is very little in the sidewall.

The deformation of the tyre owing to centrifugal fling at high speed is also shown. At 150 mph, the cross ply not only grows across its diameter, it shrinks across its shoulders. This change increases wear at the tread centre, making it flatten, which in turn produces uneven camber thrusts which can cause high speed weave.

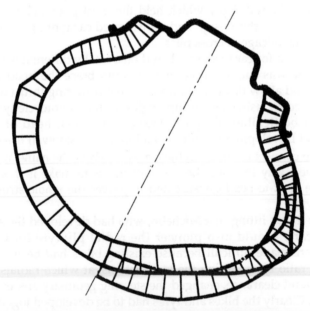

Figure 3.2 The distortion of a 160/80VR 16 radial ply tyre under load and cornering forces at a camber angle of 30°. Note that most of the deformation occurs in the crown and not in the sidewall. Lower profile (/60) tyres have still less sidewall flexing. (*Michelin*)

The 160/80, 'compromise' radial shows a similar amount of fling but much less contraction in width, while the 160/60 'fully developed' radial shows hardly any distortion at all. Apart from the ability to use softer compounds without making the tyre overheat, this offers much less weight (the saving in inertia is considerable as this is the fastest-moving thing on the bike) and wear which is as even as it can be, so the tyre's handling characteristics do not alter as much as the tread wears down.

Contact area

The contact patch depends on the original profile of the tyre and its flexibility in the crown and the shoulder. In general, the larger the area the better. It also needs to change progressively as the bike rolls, because the patch will change shape, its centre of pressure will move and as centrifugal force loads up the bike there will be more deflection at the tyre. Therefore if it rolls on to a smaller area and at the same time generates a larger force, the tyre will have to deflect, using its spring energy to support the bike. Sudden changes will cause deviations in the camber thrust and therefore in the steering effort, which the rider will feel as vague steering or as the bike moving about beneath him. The same applies to tread shuffle or squirm caused by the compression/expansion cycle.

The use of wide tyres causes another difficulty when the bike rolls because the contact patch moves further and further from the bike's centreline and from its steering axis. The taller the tyre section the more pronounced these effects are; lower sections minimize them, for a given width of tyre. There are several effects. It reduces the camber thrust for a

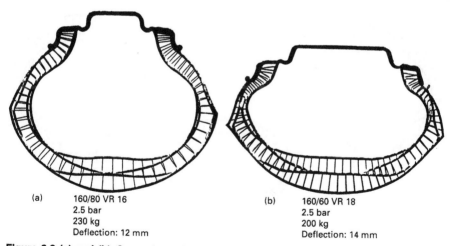

(a)   160/80 VR 16
2.5 bar
230 kg
Deflection: 12 mm

(b)   160/60 VR 18
2.5 bar
200 kg
Deflection: 14 mm

Figure 3.3 (a) and (b) Comparison of the deflection of a 160/80VR 16 'compromise' radial (a) and a 160/60VR 18 fully developed radial (b) under load. At the same pressure, the low profile tyre deflects more for slightly less load. Again, most of the deflection is in the crown of the tyre. The low profile tyre has a significantly wider contact path. Note also the change in rim width. (*Michelin*)

given angle of lean. Ironically, a bike with a higher centre of gravity or with the centre of gravity shifted in towards the turn, will require less angle of lean to balance its centrifugal force. For a front tyre, it sets up a greater self-aligning torque if the contact patch is displaced to one side. The rider would feel this as a constant pull on the handlebars as the steering tried to turn into the corner. Brake torque would increase this effect, producing steering torque and, as the bike's inertia acts through the steering axis it can create a roll couple as well. The two can oppose one another and tend to cancel one another out, or they can add together, in which case using the front brake while banked over makes the bike want to straighten up quite forcibly. This interaction depends on the trail and the front tyre profile. It can be reduced by using less trail, less castor, a front tyre which is less wide (or has a narrower contact patch) and one which has a lower section height.

The distribution of brake torque over the width of the contact area is not symmetrical when the bike is leaning over; the radius of the tyre decreases towards the inner edge of the contact area, so the force here will be greater. Braking will therefore tend to shift the centre of pressure further away from the tyre's centreline. It will also tend to turn the wheel in the direction of the corner.

A similar thing happens at the back tyre when power is used. More force is generated at the edge of the contact patch near the tyre's shoulder and the centre of pressure is moved towards this edge. The torque on the contact patch tends to turn the wheel away from the corner. The things which govern the steering effort are the castor and trail and the shape of the contact patch because this defines where the centre of pressure is. More trail moves

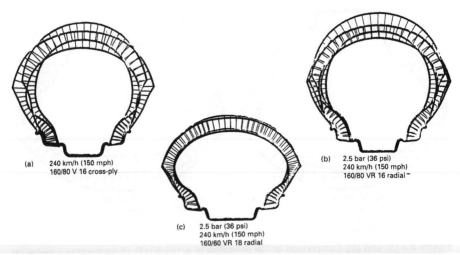

(a) 240 km/h (150 mph)
     160/80 V 16 cross-ply

(b) 2.5 bar (36 psi)
     240 km/h (150 mph)
     160/80 VR 16 radial

(c) 2.5 bar (36 psi)
     240 km/h (150 mph)
     160/60 VR 18 radial

Figure 3.4 (a), (b) and (c) Comparison of the centrifugal growth ('fling') of three tyres at high speed. The cross-ply (a) grows in the centre of the tread and contracts across the shoulders. The same size, compromise radial (b) has similar growth in the centre, but less contraction. The fully-developed radial (c) has much less deformation in all directions. (*Michelin*)

the centre of pressure further from the steering axis, so it will be displaced further from the bike's centreline when the steering is turned. Steeper castor also increases this displacement. The contact patch will be elliptical; a long, thin ellipse offers less scope for the centre of pressure to be displaced to one side, a wide ellipse is more susceptible to unequal torque distribution during braking.

The steering movement at the handlebar is also related to the castor and camber (lean) angles:

$$\alpha = \beta \cos \phi / \cos \theta$$

Where α is the steer angle at handlebar, β is the steer angle at wheel, ϕ is the camber from vertical and θ is the castor from vertical.

An example using typical values of 26° for castor and 45° for camber shows that to turn the wheel 1° when the bike is vertical requires a movement of 1.11° at the handlebar. When the bike is banked at 45°, the same 1° change in wheel steer requires 0.79°; so, as the bike is banked over the steering becomes heavier but more direct (see Figure 2.9).

When the bike is banked over to follow a curve of a particular radius, the steering will be turned a certain amount, depending upon the bike's speed. According to the castor and trail, the tyre size and the camber angle, this could put the tyre's centre of pressure to the right or the left of the steering axis, or directly in line with it. If it is to the left of the steering axis in a right turn, this will require a steady pull on the right handlebar to maintain the cornering effort. If it is to the right, it will require a steady pull on the left handlebar; if it is on the steering axis, the steering will be neutral. If it is neutral in steady conditions (i.e. all the forces balance) then it will need a balancing force at the handlebar during acceleration (when the thrust is greater than the drag) and when the rider wants to steer or alter the angle of lean.

By altering the castor, trail and tyre section, these steering characteristics can be changed to suit the rider or the conditions.

The tyre does not have perfect purchase on the ground; as it rolls it can slip, that is it can travel at a slightly different speed to the speed of the bike and it can travel at a small angle to the direction in which the wheel is pointing (slip angle). This has an effect on the steering because it alters the instant centre about which the bike is turning, as shown in Figure 3.6. A small, rigid, hard compound tyre has less capacity to generate slip angles than one which has a large contact area, and flexibility either in the tread compound, the crown or the sidewalls.

Most of the bike's cornering force comes from camber thrust, some from steer angle, elastic deformation in the tyres and some from the slip angles generated at the tyres. The combination of trail, castor and tyre profile determine the steer effort when the bike is banked over, while the lean angle needed to balance the cornering force depends on the tyre profile and the position of the centre of gravity (which the rider can change slightly by moving his

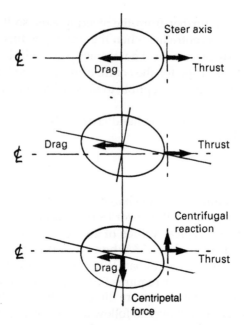

Figure 3.5 The front wheel is pulled along behind the steering axis; thrust appears at the steering axis, drag from the tyre at its centre of pressure. The two forces are in line when the wheel is straight and upright (top). When it is turned to the right, the contact patch is displaced to the left. The drag and thrust forces set up a self-aligning torque which tends to straighten the steering (middle). If the wheel is turned to the right and banked to the right, the contact patch will also move to the right (bottom). Depending on the amount of lean and the amount of turn, the centre of pressure could stay to the left of the centreline, be on the centreline or move to the right of the centreline as shown. This could create steering torque which tries to straighten up, remains neutral (zero torque) or tries to turn into the corner.

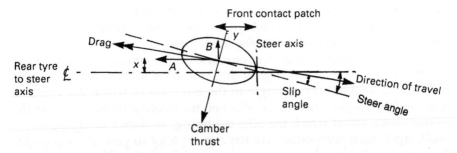

Figure 3.6 Each force acting on the tyre has a component parallel to the bike's centreline (*A*), and another at right angles to it (*B*). If the contact patch is displaced as shown (which depends on the tyre profile, the steering geometry and the lean and steer angles) then there will be a self-aligning torque *Ax*, opposed by a torque *By* which tries to move the steering further into the turn. If the total (*Ax* − *By*) is positive then the rider will need to pull on the right bar to maintain a right turn; if it is zero the steering will need no force and if it is negative, the rider will need to pull on the left bar to hold a steady right turn. Braking will increase the value of A. When the bike is leaning the vertical reaction to its weight will set up a torque about the steer axis, tending to turn the steering into the corner – see page 31.

position). The bike is said to oversteer if:

- increasing cornering force is produced by constant or lessening steer effort.
- the rear slip angle is greater than the front

Because of this it is necessary to optimize the castor, trail and centre of gravity for a given tyre profile, and to adjust the suspension to match the spring characteristics of the tyre.

As an example, imagine a bike banked into a right turn, at a steady speed and constant radius. If the steering is now turned slightly to the right the bike could increase its rate of turn, and reduce the turn radius. Or it could move further upright, reducing its cornering effort and increasing the turn radius. The result depends upon the way the bike is set up and its speed; in the first case it is said to be 'pushing' the front end and this is more likely to happen when the bike is cornering with the power off, or is being braked.

Tread pattern

It is taken for granted that the softest compound will give most grip but will overheat more easily and will wear faster. Thus, a lighter, stronger construction can run cooler and can use softer compounds. Racing 'wets' can use very soft compounds because they are water-cooled and are not used as hard as dry weather tyres – used in the dry they would quickly overheat and

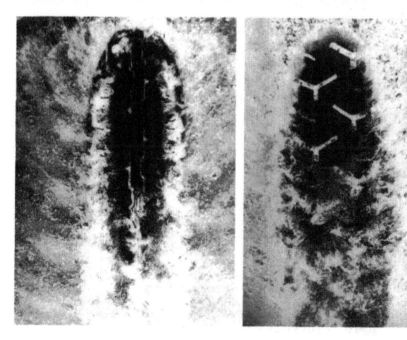

Figure 3.7 Michelin's A59X (*left*) and M59X (*right*) radial ply tyres at 80 mph in the wet, photographed through a glass plate. The shapes of the contact patches are clearly shown, as is the area of rubber-to-glass contact. (*Michelin*)

break up. But they also need to remove the water from between road and tyre, which is what the pattern is designed to do. Some manufacturers say that, within reason, the shape of the pattern is not important. Others say it is critical. Certainly, under some conditions, the compound is more important for wet grip. In wet braking tests run by *Performance Bikes* at MIRA, a very soft compound slick performed better than all of the sports compound road tyres. However, Metzeler says the pattern is important and it did a lot of testing before settling on its ME33 and this tyre in Comp K compound beat all of the other sports tyres in the same test.

The front tyre needs to have more capacity to drain away water than the rear, because it splashes water out of the path of the other tyre. Both tyre patterns need to be at their most efficient when the bike is banked 20 to 30° from the vertical, which is usually the limiting condition in the wet.

The tread pattern has another function: as the pattern moves through the contact patch, the individual strips or blocks of rubber are first deflected and then loaded with the weight, cornering force, power or brake force which the tyre is transmitting. As well as the overall deflection of the tyre, there is the possibility of an individual piece of the pattern deforming or moving. The result – often described as tread shuffle or squirm – is that the grip fluctuates, probably the centre of pressure moves around or possibly the contact patch itself moves. This is not particularly desirable for road or race tyres. It has some benefit in off-road tyres because the movement tends to prevent the pattern becoming clogged by mud.

Most sports tyres now have a pattern which consists mainly of diagonal channels which move smoothly through the contact patch with the least interruption and which leave well-supported strips of tread in contact with the road surface. The pattern is at its most comprehensive about 20° around from the centreline and the front tyre has more pattern than the rear, particularly near the centreline.

Chapter 4

Wheels and driveline

The wheels and the various things attached to them are critical to the bike's performance. The masses of these parts have rotating inertia which has to be increased or decreased whenever the bike is accelerated or braked: it is unsprung mass, and therefore governs how the suspension performs (see Chapter 5); it is steered mass which affects the steering response and creates gyroscopic forces which interact with the steering. Every pound of material carried on the wheels is worth two to five pounds carried elsewhere on the bike.

The job of the frame is to carry the wheels rigidly and keep them in the same plane. It is assumed that they are originally fitted in the same plane, if not then the steer force or the camber thrust caused by the wheels being out of line will make the tyres 'scrub', which will both wear the tyres and use up engine power.

When the bike is first set up, the wheels should be on the centreline of the frame and the drive sprockets aligned exactly (with enough clearance for the tyre). It is best to align the wheels before the tyres are fitted, so that the strings or straight edges used can run along the edges of the rims and so that the difference in front and rear rim widths can be accurately measured and allowed for.

The swing arm and, often, the rear frame may be asymmetrical and it is impossible to take reliable measurements from these parts. Instead, centre the front wheel in the forks and align the rear wheel with it, checking that the steering is dead ahead and allowing for differences in the rim sections.

Once the rear rim is correctly positioned its width (at the hub, across its bearings, etc.) can be measured and the centreline can be calculated from some fixed reference point – typically the flange which carries the sprocket or the flat, parallel portion of the swing arm where the spindle is located. Make all future measurements from this datum.

The alignment should be checked through the full range of suspension travel (minus the springs, of course, which is also a good time to measure the wheel travel against the compression of the spring strut if you want to know the leverage or 'rate' of the suspension linkage). The engine needs to be positioned so that the chain line clears the tyre – which may mean moving the engine or extending the sprocket offset. If the sprocket or gearbox shaft is extended by more than a few millimetres (particularly for very powerful drag race engines with very wide back tyres) then an outrigger bearing must be built in to support the shaft (see Figure 2.7)

Twist or bending in the frame can often be straightened, but it needs specialist jigs and the material, especially aluminium alloy, may need heat treating afterwards. The frame manufacturer is the best place to go for advice; they may also have crack-detecting dye. Small cracks and notches in aluminium and magnesium alloy parts can be TIG welded and machined back to the normal profile – it is essential that they are because any surface irregularities will cause stress concentrations which weaken the part locally and propagate the crack. Magnesium alloy parts also tend to corrode badly if the surface protection is damaged, so as well as repairing notches, etc., scratches should also be painted over promptly.

Drive chain

When the chain line has been checked, move the suspension until the chain is at its tightest point (which will be when the swing arm spindle, wheel spindle and sprocket centre are all in line). Adjust it until the chain tension is just tight at this point and then measure the chain tension when the suspension is in the normal, static position. Calibrate the old adjuster marks, or make new ones. This is the tightest the chain must ever be run; if that makes it too loose in other positions then consider fitting a tensioner, or a protective nylon strip over the swing arm, or redesigning the suspension geometry.

Chains obviously have to be strong enough to transmit the torque to the rear wheel, but they also have mass (which is rotating and most of which is unsprung). Heavier chains tend to be stronger but they are not necessarily better at power transmission: they add to the rotating inertia and at high speeds they tend to lift off the sprockets, which increases wear and tension in the rest of the chain.

For high performance use, chains need to be as light as possible (racing machines commonly use 'thin' chains and sprockets, Honda developed chains with hollow pins, Kawasaki developed cutaway sideplates). The necessary strength comes from the quality of the chain, not its physical mass. The difference between high quality and ordinary chain is in the material and the method of construction, mainly how the pins are attached to the side-plates. A poor quality chain will deform at the sideplates, making 'stiff' links, which will eventually fail. This local misalignment can also load the rollers unevenly, leading to failure.

The same symptoms appear if the wrong type of chain is used for the sprockets. Chain sizes are coded (see note) but the codes do not make it obvious when the roller dimensions have been changed.

Note: The size is expressed as the pitch (distance between pin centres) multiplied by the width (inside walls of the sideplates), as in 5/8 x 3/8 in. Chains are conventionally made to inch dimensions, specifically to units of one-eighth of an inch. The size coding reflects this, for example 530. The first digit, 5, represents the pitch, in eighths of an inch (i.e. 5/8). The next two represent the width, in eighties (i.e. 30/80 in this case). 20 would be 20/80 or 1/4 inch and 25 would be 25/80 or 5/16 inch. There are exceptions, when the manufacturers alter the final digit to denote, usually, a stronger chain. For instance a 428 would run on the same sprockets as a 425 but it has thicker sideplates. A 532 and a 632, however, have larger rollers and do not 'gear' properly with 530 or 630 sprockets. If there is any doubt, check with the supplier.

Chain lubricant

The choices here come down to extreme pressure (EP) oil, chain grease and aerosol lubricant. Where O-ring chains are used, the lubricant is mainly needed to prevent corrosion and to help the chain run quietly; make sure that any lubricants – particularly aerosols – are compatible with the seals.

Ideally, the chain (not O-ring types) should be degreased in a solvent, cleaned and dried before lubricating. In this case, chain grease is probably the best choice, warming it in a pan until it can flow into the rollers, but not letting it boil. Some of these greases are extremely sticky and may not be a good idea if the bike is used in dusty conditions.

Gear oils (extreme pressure oils) can be brushed into the chain easily and work perfectly well. The only trouble is that they can run out just as easily and need frequent applications. They are good on O-ring chains.

Aerosol lubricants vary enormously. In essence they have a grease carried in a solvent so that it can penetrate into the bushes inside the rollers and a foaming agent which helps them do this. The solvent then evaporates, leaving the heavier grease in place. Some types stay in place, others do not. Any type will do if you do not mind the mess around the back of the bike and you are prepared to lubricate the chain frequently. Products made by Motul, Silkolene and Rock Oil work very well. Some aerosol lubricants leave the chain dry to the touch – this is a useful feature on dirt bikes used in sand and dust which would otherwise stick to the chain and increase the wear rate.

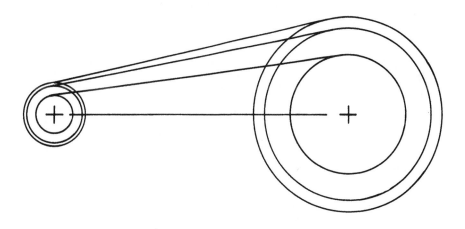

Figure 4.1 These three sprocket combinations, drawn to scale, will give exactly the same reduction (27/9, 39/13, 45/15). Compare the difference in sprocket size, chainline angle and clearance. For the same gearbox torque, the force in the chain will be 1.67 (i.e. 15/9) times greater if a 9-tooth sprocket is used instead of a 15-tooth

Sprockets

The relative size of the sprockets determines the final reduction gear (= t_w/t_g, where $t_{w, g}$ is the number of teeth on the wheel sprocket and gearbox sprocket respectively. To raise the gearing, i.e. turn the back wheel faster for a given engine speed, it is necessary to reduce t_w or to increase t_g). There are other effects.

Larger sprockets will obviously be heavier and will have much greater inertia. A certain minimum diameter is necessary for the chain to clear the swing arm, while a front sprocket which is too large may make the chain touch the gearbox/crankcase casting. Very large rear sprockets, on trials bikes for example, can cause clearance problems: touching rocks, picking up mud, entangling undergrowth, string or wire. The effort to reduce the size of the rear sprocket while keeping the same gear ratio has led to the use of gearbox sprockets as small as nine teeth. The smallest practical size for higher outputs and wheel speeds is thirteen teeth.

Changing the sprocket size, even if the other sprocket is altered to keep the same gear ratio, alters the force in the chain and the angle of the chainline (see anti-squat suspension, Chapter 5). If the torque at the gearbox output shaft is T, then the force in the chain will be T/t_g. Therefore the larger the gearbox sprocket, the smaller the force in the chain. The centrifugal force – which tends to lift the chain off the sprocket – increases in proportion to the radius of the sprocket and to its rotational speed squared.

While final drive sprockets of 39/13, 42/14 and 45/15 will all give exactly the same gear ratio (see note), there will be other differences which will make one pair more suitable than the others, depending on the use to which the machine is put (whether the most important factor is clearance, chain strength, unsprung weight or the effect of chainline force on the suspension).

Note: The overall reduction ratio between crankshaft and rear wheel is $R = pGf$, where p is the primary reduction = t_{cl}/t_{cr} ($t_{cl, cr}$ – teeth on clutch and crankshaft sprockets); G is the internal gear ratio selected = t_o/t_i (teeth on output and input shaft gears); f is the final reduction (t_w/t_g). Note that if there are countershafts, bevel gears, etc., f will have to be calculated accordingly. The same applies to p and to G if a layshaft type of gearbox is used. The reduction ratio is the size of the driven gear divided by that of the driving gear. This is a reduction gear, so the speed of the rear wheel will be engine speed (n) divided by the gear (= n/R) while the rear wheel torque will be crankshaft torque multiplied by the gear ratio (= crank torque x R, ignoring friction losses).

If n is in revolutions per minute and if the rolling radius of the tyre is r inches, then the speed of the wheel (bike) in miles per hour is 0.00595 $nr/R = nr/168.07 R$.

There are several ways to reduce the weight of the larger rear sprocket: to use lighter materials (aluminium alloy is common, plastics appear every so often); to reduce the width of the sprocket (which must increase its wear rate); to drill the sprocket.

Sprocket carriers

High tensile bolts, fully tightened and locked (using tab washers or wire) must be used for rear sprockets and the plain shanks of the bolt must be a good fit in the hole in the sprocket. If some kind of cush drive can be incorporated then use it, because it will protect the gearbox, clutch and crankshaft gears.

Wheels

The wheels are dictated by the choice of tyres (see Chapter 3) but having established the size, it is necessary to keep the weight down to the absolute minimum. The inertia of a rotating part is proportional to its mass and to the square of its radius. The mass of a wheel is concentrated at the rim, at the largest possible radius.

Some people choose the wheels first, only to find that the desired tyres are not available. The reason is that high performance tyres have been developed very rapidly in recent years, for a small number of machines, so there has been neither the time nor the need to produce a full size range. The latest developments will only be available in a few sizes and the wheels have to be selected to take advantage of this.

The choices are between wire-spoked wheels and cast wheels, and then between the various materials available. Unless flexibility and rapid repairability are important – as they are on dirt bikes – then the rigidity of cast wheels gives noticeably better handling.

As the size of the wheel is the same, then materials of less density are the obvious way to save weight and increase stiffness (see Chapter 8). Experiments with plastics (on bicycles) and carbon fibre show the way to go. Proven materials are cast alloys of aluminium (cheapest and heaviest) or magnesium, which has two-thirds the weight. In between there are sheet alloy pressings and rivetted structures which combine lightness with low cost.

Wide rims coupled with low tyre sections make it increasingly difficult to use tubes (and, therefore, wire-spoked wheels); BMW (Akront) and Honda have both produced wire wheels which will take tubeless tyres for off-road machines.

On a competition bike the chassis will have been lightened drastically and the unsprung mass reduced as much as possible but it will inevitably have a worse sprung/unsprung ratio than a road bike. This means that the suspension will not work as well as it might (see Chapter 5) and that it will be more sensitive to any additional force, created by the wheels being out-of-round or out of balance. The tolerance for runout at the rim should be zero; the slower and the heavier the bike, the more it can tolerate.

Static wheel balance is generally considered adequate for bike wheels but the width of the rims used on some racers, and certainly on drag bikes, means that dynamic balancing ought to be used. The difference is shown in Figure 4.3. In the absence of a dynamic balancing machine, balance weights should be added as close to the wheel centreline as possible, or they should be divided equally on either side of the centreline. The balance should be checked when a new tyre has been used for 10–15 minutes, or register marks should be made on the tyre and rim to confirm that there has been no movement.

Gyroscopic effects

The inertia of a wheel – which is proportional to its mass, the radius (squared) at which the mass is concentrated and its speed – shows up in

Figure 4.2 Light, rigid wheels are essential for good handling and good suspension, particularly when the sprung part of the machine has been extensively lightened. Astralite's rivetted sheet alloy construction is one option (**a**). Their solid prototype (**b**) offers a potentially lighter construction for the same rigidity. Maximum rigidity for minimum weight comes with the more expensive cast magnesium alloy wheels, such as this 17-inch Maxton (**c**). The cush drive is from a Honda RS250

gyroscopic forces. One of these is called precession, the ability to translate a force into a different plane. This may be demonstrated by holding a bicycle wheel by its spindle and spinning it quite rapidly in the 'forward' direction. Try to *turn* the wheel to the left and it will respond by *leaning* equally forcefully to the right. The torque applied to the spindle is shifted 90° in the direction of the wheel's rotation.

This is quite useful during the initial steering phase because, to make the machine bank to the right, the rider tries to turn the steering left. Precession at the front wheel helps the steering effort.

Now as the whole machine begins to lean to the right, precession at the rear wheel, at the crankshaft and the gear shafts, creates forces which make those parts want to *turn* to the right. Or to the left if, for example, the engine is turning 'backwards'. This sets up bending forces in the swing arm and engine mounts. Apply a bending force to something and it will bend, there is no question about that; the only question is, how much? Even slight bending at the rear suspension and main frame can alter the wheel alignment enough to change the cornering attitude. And when the bike has settled on its course, the steering forces are removed and the gyroscopic reaction disappears so that the bent frame springs back to its normal position, changing the wheel alignment yet again. Therefore a reduction in rotating inertia and an increase in frame/swing arm stiffness are both very desirable.

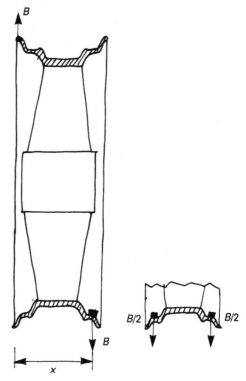

Figure 4.3 If an out of balance force (B) is near the edge of a wide wheel and the balance weight is added to the opposite edge, the wheel will still balance statically but there will be a rocking couple *Bx* created. This could be halved by dividing the balance weights evenly between each side of the rim or by putting them close to the centre

Wheel bearings
Use a wheel bearing grease (a water repellent type for dirt bikes) and inspect and repack the bearings frequently. The bearings should be spaced accurately, both internally and between the bearing and the fork leg, using ground flat spacers. When the spindle and its clamps are tightened up, there should be no preload on the bearings. If the spacers are not dimensionally accurate, or if the wrong tightening sequence is followed, it is possible to distort the leg enough to make a disc brake drag. Slotted axle nuts should also be shimmed rather than overtightened to make the slots line up with the hole in the axle.

51

Chapter 5

Suspension

The first requirement for suspension is to cope with bumps: partly to prevent jolts from the road making the rider uncomfortable and damaging parts of the bike, and partly to keep the tyres in even contact with the ground, so that they have constant grip for acceleration, braking and cornering.

The only difficulty with this is that we do not know how big these bumps are. Imagine a bump which climbs a total height h over a distance x (and whose edges are generously radiused, to lead a wheel gradually onto the upward slope). A bike with no suspension, travelling at speed v, takes time $t = x/v$ to go from bottom to top. During this time it has an average vertical speed of $h/t = hv/x$, and it has to accelerate from zero (vertically) to this speed. The most time it has is t, which would give it an average acceleration of $hv/xt = hv^2/x^2$. This is the lowest possible figure (if it took longer than time t to accelerate, it would have gone straight past the bump), but already we can see that the acceleration (and therefore the force needed) is proportional to the height of the bump, proportional to the bike's forward velocity squared and is inversely proportional to the horizontal length of the bump, also squared.

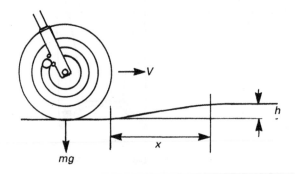

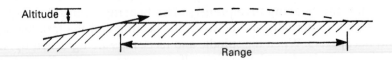

Figure 5.1 Bump dimensions and the trajectory produced by a bump

But what is happening during the time interval that the bike takes to accelerate vertically? If the acceleration (and the force) is allowed to build up gradually, what delays it? We could imagine the tyre hitting the edge of the bump, compressing and then passing on the force to the rest of the bike. But what if the tyre were rigid? The vertical speed would go from zero to maximum in the time it took the wheel to roll from the horizontal on to the slope of the bump, i.e., t would tend to something very small indeed, and the expression hv/xt would therefore tend towards something very large indeed. This is the bike's acceleration, and the force needed to produce it is equal to mass times acceleration, so the bump force, on a rigid wheel, is immense. (The alternative is that the rim would deflect – buckle – and absorb some of the energy this way.) Let us allow some flexibility somewhere and assume that it gets halfway up the slope before it reaches its full vertical velocity, i.e., the time taken is $t/2$ (= $x/2v$). Then its acceleration will be $2hv^2/x^2$, something which is heavily laden with assumptions but it is a starting point from which we can produce some numbers which will at least show the order of magnitude of the problem. Table 5.1 shows a variety of speeds over a bump which is 4 feet in length and just half an inch high, and the acceleration they produce at the wheel.

Table 5.1 Bump force versus speed

Bump height:	0.5 in	0.042 ft			
Bump length:	4 ft				
Speed v, mph:	20	40	60	80	100
ft/s	29.3	58.7	88.0	117.3	146.7
Acceleration:					
$2\,hv^2/x^2$, ft/s^2	4.46	17.9	40.3	71.5	113.0
g	0.14	0.56	1.25	2.22	3.51

So the same fairly small bump would generate a force in the region of half the bike's weight at 40 mph, over twice the bike's weight at 80 mph and three-and-a-half times its weight at 100 mph. This is allowing for some flexibility in, say, the tyre, which prevents the acceleration from becoming instantaneous, and spreads it over a longer period, thus cushioning the shock.

This is the prime task of the suspension. The acceleration (and therefore the force) generated by the bump is:

$$a = 2\,vh/xt \qquad\qquad (5.1)$$

Where a is the vertical acceleration due to the bump, v is the horizontal velocity, h is the height of the bump, x is horizontal length of bump, and t is the time taken for the bike to accelerate to full vertical speed.

Now if we wanted the bike to be able to go over the bump in the previous example, at 100 mph but without exceeding a force of 2 g, the requirement for the suspension is fairly simple:

maximum acceleration $2\,g = 2vh/xt$

therefore, $t = vh/xg$

The suspension has to spread the load over this period of time during which the bike would travel a distance of v^2h/xg and if this is longer than x, the length of the bump, then the requirement is not physically possible. It would be possible to calculate the limiting speed for the bump. However, there is another way: to let the wheel take the full bump force and to delay the transmission of this force to the rest of the bike.

As the tyre first encounters the bump, there will still be some deflection and then the wheel will be accelerated upwards, this time compressing the spring in the suspension. (The tyre plays an important role here and the suspension needs to be adjusted to suit the spring characteristics of the tyre.) As before, the acceleration will be $2\,vh/xt$ but now it is only applied to the unsprung mass of the wheel, brake and lower suspension leg. The force is much smaller and compresses the spring a corresponding amount, setting up an equal force in the spring which immediately begins to lift the bike. First, the time interval has been extended, from merely compressing the tyre to deflecting both the tyre and the spring. Second, the force transmitted to the bike is the force stored in the spring. This is $m_1\,2\,vh/xt$, where m_1 is the unsprung mass of the wheel. Reduce m_1 and you can reduce the force still further, *regardless of the amount of mass in the rest of the bike.*

Assuming the rest of the bike is extremely heavy and has not moved appreciably in this time, the force would compress the spring a certain distance ($= m_1\,2\,vh/xts$, where s is the spring rate). If this is less than the height of the bump then the spring is too hard, if it is more than the height of the bump, then the spring is too soft – *at that particular speed.*

The spring force acting on the sprung mass, m_2, will move it upward with an acceleration a_2:

$$a_2 = m_1\,2\,vh/xtm_2 \qquad\qquad (5.2)$$

This is the same acceleration equation as before, but modified by the ratio m_1/m_2, the ratio of unsprung to sprung mass and the smaller m_1/m_2 can be, the smaller will be the acceleration of the main part of the bike. This is why it is fairly easy to engineer comfortable suspension for heavy vehicles. For high performance vehicles it is necessary to reduce m_2 as much as possible and it then becomes increasingly difficult to reduce the unsprung mass in proportion (or ideally, even more). This explains the attraction of magnesium alloy wheels and brakes, carbon fibre wheels and discs, light tyre construction, etc. There is not much that can be lightened on the unsprung side of the suspension, and the only way to get large gains is to use exotic materials. Upside-down forks, with the stanchions carrying the wheel and the larger sliders, complete with damper construction and oil, attached to the sprung mass, have become the norm on GP racers. Note also that while the bump force on the sprung part of the bike is reduced considerably by the suspension, the unsprung part – mainly the wheel – has to carry the full bump force and must be strong enough to cope with this. The rotation of the wheel puts it in fatigue, so the maximum bump force must not exceed the endurance limit of the material. The survival criterion for fatigue tests is usually 6 million cycles; for a typical road wheel, 10 million cycles represents a distance of about 12,000 miles.

So far the suspension can be seen working in two ways, first to let the bump force act on a relatively light mass and to transmit this smaller force to the rest of the bike; second, to spread the shock load over a greater length of time. The unsprung mass also gains inertia, which would eventually be fed into the spring and passed on to the rest of the bike. This can be countered by using compression damping, which is any force that can be used to oppose the spring's compression. In practice it is achieved by forcing oil through small passages, which makes the oil hot and dissipates energy which would otherwise be transmitted by the spring. This force is added to the spring force and effectively makes the suspension stiffer but its main object is to prevent the unsprung mass from reaching too high a speed and gaining inertia which would make it more difficult to control. The pressure in the fluid moving through the damper orifice is proportional to its speed squared, so the damping force is highly speed-sensitive. This becomes quite useful in the next phase of operation. We left the bike with its front wheel rushing up the slope of a bump, transmitting a now reduced force to the bike, which is accelerating upwards at a much lower rate than the wheel. As it reaches the top of the bump, the suspension has compressed by an amount h if the spring rate is right, possibly by more if the spring is too soft. The wheel also has upward momentum and wants to keep travelling, even though the bump has dropped away and is no longer in contact with it.

We now have the bike travelling upwards at a low speed, the wheel travelling upwards at a higher speed and the spring trying the push the two apart.

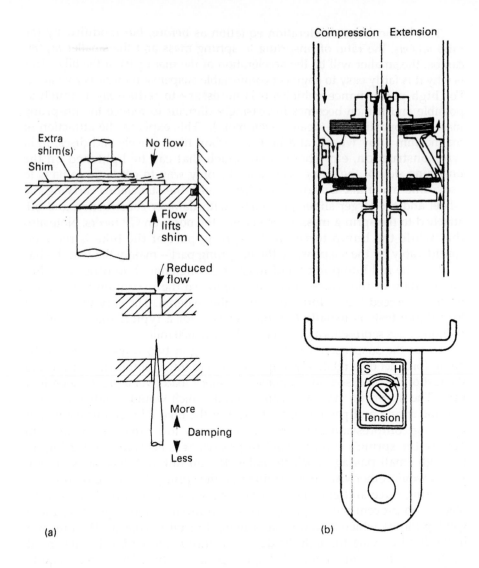

Figure 5.2 **(a)** Damper mechanisms. *Top*: a piston, sealed by a piston ring, is pushed up and down in a chamber of oil. Holes covered by sprung shims allow passage of oil in one direction (and produce a force in proportion to the size of the hole and the strength of the shim). This also acts as a one-way valve. *Centre*: using the shim to block part of the hole creates a small resistance in one direction and a much larger force in the opposite direction. *Bottom*: using a tapered needle to restrict the hole provides a variable damping force which can be altered either by an external adjustment or when parts of the damper move relative to one another. **(b)** The rear damper unit used in the VFR400R Honda, showing the oil path during extension and compression and the needle adjustment. (*Honda*)

On its own, the wheel would now follow a ballistic course, it would take off in a neat parabola and land some distance down range of the bump. If the up-slope of the bump was at an angle α (so that tan α = h/x) then the unrestrained wheel would fly on this course:

maximum height $= (v^2 \sin^2 \alpha)/2g$

range $= (v^2 \sin 2\alpha)/g$

flight time $= (2\,v \sin \alpha)/g$

With solid suspension, or given a slope long enough for the suspension not to regard it as a bump, the entire bike would follow the same course, although air resistance would slow it down and reduce the above values slightly.

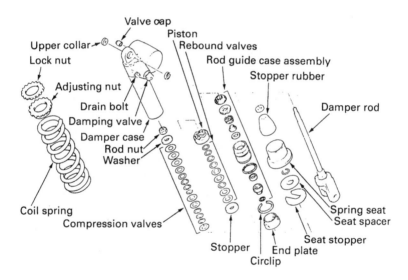

Figure 5.3 Component parts of the rear damper used in the Honda CR500R motocrosser. (*Honda*)

However, part of the suspension's job is to keep the wheels in contact with the ground. At the top of the bump, where the tyre would try to leave the ground, it has the spring pushing it back down. The bike also has the spring pushing it up. As the unsprung mass is smaller, the spring tends to make it react faster, accelerating the wheel downwards, although the sprung mass of the bike will still be pushed upwards to some extent. The ratio between the two masses will determine the relative motion; the lighter the unsprung mass, the easier it will be for the spring to move it and not the bike.

The bike now follows its own trajectory, hopefully with the suspension extended so that the wheel is in contact with the ground. As the full weight of the bike returns to the spring, it will compress, probably go past the static point and bounce back. This is all gratuitous movement which is unwanted and is reduced by having quite strong extension (or rebound) damping.

To allow for extension movement, i.e. if the bump began at the top and had a downhill slope, the static position of the suspension needs to be some way between the full bump and fully extended positions, usually one-third of the travel is used. The spring rate (or rather the wheel rate) is determined by the ratio of the unsprung to sprung mass and the likely bump forces involved at a particular circuit. The optimum is the softest possible spring, which bottoms occasionally but infrequently. Note that as the speed increases, so does the bump force.

For road bikes, the designer will settle on an arbitrary maximum acceleration for the sprung part of the bike (a_2), based on previous experience and road tests, and work out a similar value for the relationship between v, h and x (the bike speed and the assumed dimensions of the bump). The biggest force that is to be expected will make the suspension bottom, and this sets the wheel rate, which is the total force at the wheel divided by the distance it has to travel. Where the springs work directly on the wheel, the spring rate will be the wheel rate divided by the number of springs. If the springs are operated via any kind of lever or are set at an angle to wheel travel then the leverage needs to be worked out either by measuring it directly from the bike or by making a scale drawing. If allowance is made for this, then starting-point spring rates can be calculated from other similar machines or from the original, unmodified machine.

Note that the spring passes on whatever force is fed into it, so the spring rate does not affect ride, it only affects how much spring travel is used up by a given bump. If you reduce the unsprung mass, it will require less force to move it over a bump, so less force will be fed into the spring (which will compress less, or you can fit a weaker spring so that it compresses the same as before) and then this force will be passed on to the sprung part of the vehicle – but it will be the same force, regardless of the spring rate and not counting anything that compression damping might have done to it. Think about this – it can make it very difficult for a rider to distinguish what the suspension is doing.

To improve ride it is necessary to reduce unsprung weight or to alter the compression damping. From this point of view there is no best or worst spring rate. A very soft spring will use up its travel too easily and will compress more than the height of the bump, which is wasteful, and will alter the attitude of the bike more than is necessary. A very hard spring will not compress enough and will pass on the shock with the minimum of delay. In addition, the reduced movement will not generate so much damping force. In between the two extremes there will be a number of spring rates which will provide various amounts of movement, some of which will be more useful than others. The weight of the bike and rider compresses springs to some extent – where they settle in the static position is called the ride height and this is controlled by the fitted load or pre-load of the spring, its length and the spring rate. The pre-load is the force which has to be overcome before the spring moves at all and is simply a means of saving suspension travel, although it does provide a useful adjustment to alter the ride height and the attitude of the machine.

When a bike meets a bump, obviously the front wheel hits it first and, if the suspension doesn't completely absorb the deflection, then the whole bike will, to some extent, be tipped nose up. This action could have an influence on the rear suspension, before the bump actually arrives at the rear wheel.

What happens is unpredictable and it can cause misleading symptoms. If

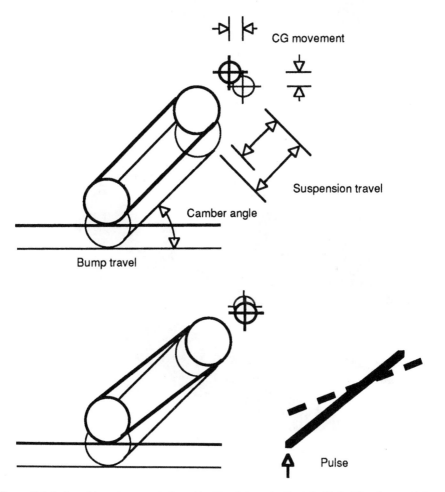

Figure 5.4 A sharp blow at one end of an object tends to make it rotate (dotted line, bottom right). This applies to a bump hitting the front wheel, tending to lift the front of the machine and possibly depress the rear. Also if the wheel is banked over, a bump can alter the angle of bank, compress the suspension or raise the whole machine – or some combination of all three

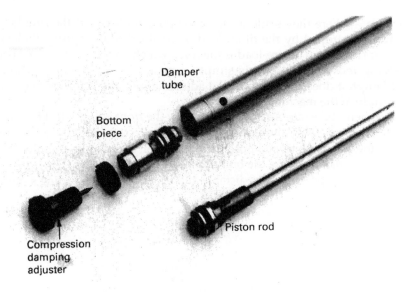

Figure 5.5

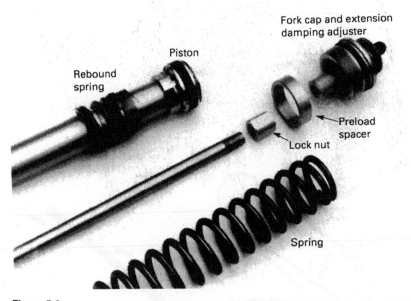

Figure 5.6

you flick one end of a pencil then the pencil as a whole will tend to move away from your finger and it will also rotate. The total movement depends on the impulse of the force, the direction of the force, its position and angle relative to the pencil and the pencil's moment of inertia about that point. It is likely that if

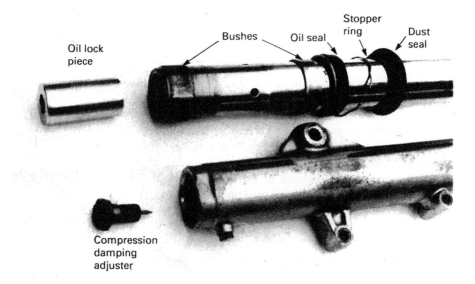

Figure 5.7

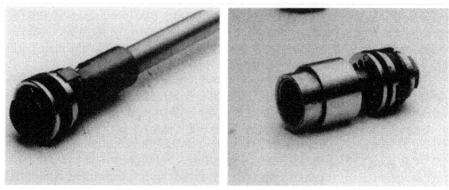

Figure 5.8   **Figure 5.9**

Figures 5.5 to 5.9 The Showa front forks used on the CR500R give 12 in of travel. During compression the damper rod forces oil through the bottom piece, which has a tapered needle adjuster. The oil lock tube at the bottom is there to make an hydraulic lock to prevent metal-to-metal contact if the forks bottom. During extension, oil is forced through the piston tube and up through the hollow damper rod, which has an adjuster fitted into the fork cap. Specialists can rebuild the damper, tailoring it by re-valving and changing the spring shims, improving it by tightening up the clearances at the bushes and the piston/piston ring assemblies

one end of the pencil moves 'up' then the other end will move 'down' and consequently there will be a point somewhere along its length which doesn't move at all.

Now, if the bump flicks the front of the bike up, a similar thing will happen to it and, depending on the nature of the bump impulse (or the surplus that

61

hasn't been absorbed by the front suspension) and the inertia of the bike, this could result in the back of the bike being pushed downwards or being raised (or not moving at all). This could leave the rear spring slightly compressed or extended (or unaffected) immediately before the bump arrives to compress it further.

If the rear spring is compressed and begins to generate bump damping force then, when it has to cope with the real bump, it will appear to be too hard. This would obviously be made worse if the front suspension had too much bump damping, for example, and therefore tended to make the whole bike lift at the front when it hit the bump. Yet to the rider it would seem as if the rear suspension was too hard.

In the same way, the back of the bike will be flicked up by the bump, tending to rotate the machine forward on to its nose, creating a pitching motion which might well seem to the rider to indicate that the front was too softly sprung.

The suspension also has to cope with bumps while the machine is banked over: as shown in Figure 5.4, this will incorporate some element of the bike being rotated (banked further) by the bump, which is potentially dangerous because it increases the angle of lean between the tyre and the road surface. While the wheel is on the uphill side of the bump this isn't so much of a problem because the contact between the tyre and the ground will be increased and the tyre will have more traction. But when the bump levels off or the wheel reaches the downhill side then the tyre loading will be reduced, just as the lean angle has reached a maximum.

The suspension can cope with bumps while banked over in several ways, as the diagram shows. It can lift the whole bike, it can compress and it can let the wheel lift over the bump while the top of the bike stays at the same height or is even tipped further into the corner. The suspension doesn't need to compress the full height of the bump in order to keep the CG at the same height relative to the bike; some combination of all three types of movement will leave the CG pretty well in its original position, which is presumably the requirement for the best ride quality.

This will not only be affected by bump damping and the spring rate but also by the moment of inertia of the bike about its wheelbase, so changes in the dimensions and weight distribution of the bike – including the rider's position – could have as much effect as suspension settings on the bike's ability to deal with bumps while cornering.

In addition to coping with bumps, the suspension also has to manage cornering forces and weight transfer during braking and acceleration. Weight transfer is severe, in the order of 100 per cent for both the front and rear suspension. To cope with such big changes, the suspension is often made load sensitive, given a rising rate, or brake/engine torque is used to augment the suspension.

Load sensitive suspension and rising rate suspension increase the wheel rate as the suspension compresses. This allows soft initial suspension but prevents bigger loads from making the spring bottom. Brake torque or a brake-sensing mechanism are used to prevent front suspension travel being used up during braking (see Chapter 6, Anti-dive brakes), while engine thrust is used to support the rear suspension during acceleration (see anti-squat suspension, below). Finally, when a bike rolls in a corner, centrifugal force is added to its own weight in the suspension. The total force on the bike, acting through its centreline is:

$$m\sqrt{(g^2 + v^4/r^2)}$$

Where m is the mass of bike and rider, g is the acceleration due to gravity, v is the velocity of bike and r is the radius of turn. The angle of lean is θ from the vertical where $\tan \theta = v^2/rg$.

The angle θ is measured between a vertical line and the line which joins the bike's centre of gravity and the centre of pressure of the tyre's contact patch. This is not quite the same as the bike's centreline and varies more from it when wider tyre sections are used, when the bike has a low centre of gravity or when the rider shifts his body weight to one side.

These forces, and the resulting deflection at the suspension, also need to be considered when choosing the spring rate. The position of the centre of gravity is moved when the suspension compresses and this can be critical for maximum traction (see Chapter 6, Anti-dive brakes), while if the bike is lowered during cornering, it may use up its ground clearance. At the same time, the suspension also may have to cope with bumps which could use up the remaining travel (or more).

Springs

Several types are used:

1   Coil

2   Torsion bar.

3   Gas.

4   Rubber in compression, tension, shear or torsion.

Coil springs and torsion bars
Steel wire coil springs are simply torsion bars fitted into a more convenient space. The wire twists as the spring is compressed and this is the elasticity which provides the spring force.

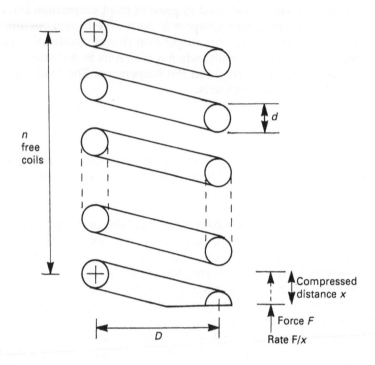

n free coils

d

Compressed distance x

Force F

Rate F/x

D

Figure 5.10 Relevant dimensions of a coil spring, the spring rate can be calculated from these measurements

The spring rate s is:

$s = F/x$

$= Gd^4/8\,nD^3$

Where F is the axial force, x is the axial distance the spring is compressed, G is the shear modulus for the material (78,500 to 81,400 N/mm^2, 11.4 ~ 11.8 × 10^6 lbf/in^2 for cold-formed to hot-formed steel), d is the wire diameter, n is the number of active coils and D is the mean diameter of the coil.

For a torsion bar:

$s = Fa/\alpha$

$= G\pi d^4/1834L$

Where Fa is the twisting couple, α is the angle through which the bar is twisted, and L is the length of the bar.

Coil springs can be used in series with one another, giving a combined rate until one goes coilbound and then giving the rate of the remaining spring. The combined rate s, is:

$$s = s_1 s_2 / (s_1 + s_2)$$

Where s_1 and s_2 are the individual spring rates. If the springs are used in parallel (telescopic forks, twin shock swing arms) then the two rates are simply added together.

An alternative to this is the dual rate spring which is wound so that the last m coils are at a shorter pitch. As the spring is compressed, these coils touch, become coilbound and no longer contribute to the spring force. The spring rate is now $Gd^4 / 8(n - m)D^3$, which has to be a higher rate than before.

There are progressively wound springs, where the pitch of the last coil is the smallest and successive pitches become greater, so that the spring goes coilbound one coil at a time, effectively giving a rising spring rate. Coil springs may also be combined with gas springs ('air-assisted' suspension), in which case the two spring rates act in parallel and are simply added to one another to get the total rate.

Gas springs
Air, or any other gas (nitrogen is the most popular), forms a very light spring which has a naturally rising rate, is infinitely adjustable and has no inertia. Telescopic forks, as long as they have efficient seals, are air-assisted even if they run at zero pressure or have no valve with which to alter the pressure. Other types have no steel spring at all and rely entirely on gas pressure. Certain types of dampers (those built on the de Carbon principle, for example) contain gas at high pressure which acts on the area of the piston rod and contributes a small amount to the total spring force.

In essence, the suspension operates a piston in a sealed chamber (which may contain oil as both a damping medium and a means of adjusting the compression ratio). In an isothermal case, i.e. where temperature remains constant, the pressure is inversely proportional to the volume:

$$pV = K$$

$$F = pA$$

Where p is the pressure, V is the volume, K is a constant, F is the force in the spring strut, and A is the area of piston.

Therefore if you halve the volume, you double the pressure. Halve the volume again (which only takes one-quarter of the stroke this time) and you quadruple the original pressure. This generates a very steeply rising rate. However, in the real world, the compression happens fast enough to cause a rise in temperature, which upsets the calculation slightly (changes in ambient temperature can also affect the 'spring' settings). The suspension just about moves fast enough for the compression to be adiabatic, that is without heat loss, and the expression becomes

$$pV^y = K$$

Where $y = C_p/C_v$, the ratio of specific heats of the gas at constant pressure and constant volume, which, for both air and nitrogen, is 1.40.

The adjustments are to the start pressure and the compression ratio (usually by adding or subtracting oil) and they have quite different effects, as the graph in Figure 5.11 shows.

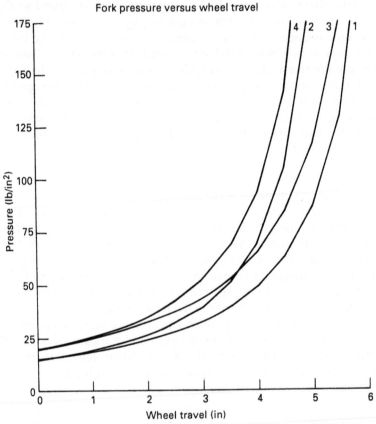

Fork pressure versus wheel travel

Figure 5.11 The relationship between gas pressure and stroke for a telescopic fork or a gas-filled spring strut. It is adjustable by altering the gas pressure or by changing the compression ratio (usually by adding/removing fork oil).
Curve 1 shows the original setting, starting at 15 lb/in² (atmospheric pressure).
Curve 2 shows what happens if the original gas volume is reduced by 14 per cent by adding more fork oil – the initial pressures do not change much but the pressure increases sharply when the forks are compressed more than half way.
Curve 3 has the same oil level as curve 1, but this time the start pressure has been raised by 5 lb/in². This gives a more gradual rate of pressure change – a harder spring than curve 2 up to just over half stroke, beyond which curve 2 gives an increasingly stiffer spring rate. Curve 4 shows the effect of making both of the adjustments – reducing gas volume by 14 per cent and raising the initial gas pressure by 5 lb/in².

66

The disadvantages are the sealing difficulties and the cost: the fact that if gas pressure is used to force seals into firmer contact then it also increases friction and wear, the sensitivity to temperature changes (especially close to an engine or exhaust system) and in being able to use the necessarily limited portion of the rising rate curve (at low deflections the rate tends to be too soft while at high deflections it becomes too high to be of any value).

Rubber springs
In the past, rubber in shear bushes have been used in pivoted-link suspension. Hagon grass track forks use rubber in tension, while early Minis had rubber in compression. The material is not used in today's bike design so there is no point in dwelling upon it except to say that it has a number of advantages, such as light weight, inherent damping and it can occupy a small space.

Disadvantages include its deterioration with age and with exposure to ultra-violet light, sensitivity to temperature and possibly to some chemicals, and the difficulty of combining a required spring force with the necessary amount of movement – although suspension linkages can solve this problem. The rubber can also take a set if it is stored under load.

Active suspension
All springs are 'passive' – they must be deflected by a force before they store any energy. An alternative type of suspension, so far only used on cars, has no springs but supports the vehicle by using fluid under pressure. Thus it needs a pumping system and a microprocessor to control it, a system which is currently too bulky to contemplate on bikes. However it can anticipate bumps and weight transfer and can control the vehicle according to a pre-set program – to keep it level, to produce positive or negative roll in corners, to prevent pitching under braking or acceleration, and so on. The response is already good enough; as component size is reduced it may be able to improve bike suspension, especially if electronic controls already have to be used for fuel injection, anti-lock brakes and anti-spin drivelines.

Several GP racers have servo-motors which can alter the spring pre-load or the damping mechanism, either controlled by the rider or operated via sensors on the bike to detect specific conditions, e.g. braking.

Damping
Early machines had friction dampers which were not successful because they produced the greatest resistive force before they started moving and, the faster they moved, the less force they produced. Hydraulic dampers, in which fluid is forced through an orifice, work the other way round. They are speed sensitive and not load sensitive.

Bernoulli's theorem states that:

$$\frac{v^2}{2} + \int \frac{dp}{D} + hg = K$$

Where v is the fluid velocity, p is the fluid pressure, D is the fluid density, h is the height above some datum, and K is a constant.

If, in an incompressible liquid, the density is constant, this can be simplified to:

$$v^2/2 + p/D + hg = K$$

The pressure is created by the force on the damper rod and the area of the piston pushing against the fluid. It does change height slightly, but the most important variable is v^2, i.e., the force varies with the square of the velocity. So the hydraulic damper is actually speed-squared sensitive.

This means that the damping force will be inversely proportional to the area of the orifice through which the fluid is pumped because the smaller the orifice, the larger the fluid velocity will be.

Fluid motion can be controlled by one-way valves, so that damping in one direction is much greater than in the other. The force can be varied by using pressure-controlled valves, for example a sprung shim which lifts at a predetermined pressure, uncovering a greater orifice to prevent the damping force building up to an unwanted level. It can also be controlled by using a tapered needle in an orifice – the further down the taper, the smaller the available area for fluid flow. This is used both as an external adjustment – usually a screw which controls the height of the needle – or to make the damping progressive and load-sensitive (as the spring unit compresses under a greater load, it moves the damper to a different position on the needle, which can give more or less damping, whichever is required).

The damping mechanism can also be controlled by opening or closing extra passages, either as a method of adjustment, or as the unit compresses, or to provide additional damping when the brake is applied (see Chapter 6, anti-dive brakes).

Damper fluid is usually purpose-made fork oil or ATF (automatic transmission fluid) but engine oil is occasionally specified. Its main function is to keep a constant viscosity over the working range of the damper but it also has to lubricate the moving parts (especially load-bearing bushes), be compatible with the seals and resist aeration.

Fork oils are available in viscosity ratings from SAE (Society of Automotive Engineers) 0W upwards. Some manufacturers list things like SAE 7½, which does not exist on the SAE scale but which is, in viscosity, midway between 5W and 10W. It is possible to blend different grades; equal quantities of SAE 5W and 10W will produce '7½', while adding more SAE 5W to it will reduce the viscosity in proportion. Don't mix different brands as they may be built to slightly different viscosities. One which meets the lower rating for an SAE 10W may be less viscous than one which meets the upper requirement for an SAE 5W!

The SAE standard requires an SAE 5W to have a maximum viscosity of 3500 cP at −30°C while an SAE 10W must have a maximum viscosity of 3500

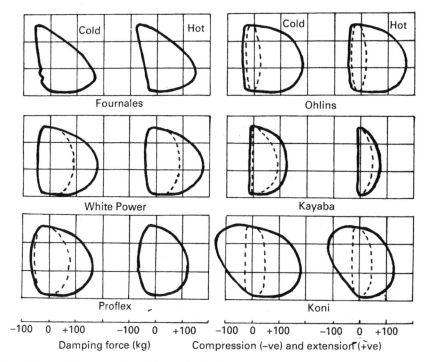

Cold Hot Cold Hot

Fournales Ohlins

White Power Kayaba

Proflex Koni

−100 0 +100   −100 0 +100   −100 0 +100   −100 0 +100
Damping force (kg)   Compression (−ve) and extension (+ve)

Figure 5.12 (a) Damper units tested on a shock dynamometer, which cycles them at a fixed speed and measures the damping force. The reduction in force after three and a half minutes running time can be seen ('hot') as can the full range of damper adjustment (dotted line shows minimum adjustment, solid line shows maximum adjustment)

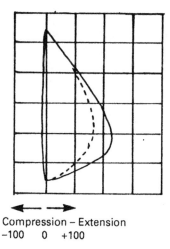

Compression − Extension
−100   0   +100

Figure 5.12 (b) The effect on damping force of changing from an SAE 5 (broken line) to an SAE 10 fork oil. The vertical scale is displacement by an eccentric shaft and connecting rod, used to pump the damper. Therefore it represents zero speed at either end of the scale, with maximum speed somewhere near the centre

cP at $-20°$C. The minimum viscosity at $100°$C is 3.8 cSt (5W) and 4.1 cSt (10W).

The way the damper works is to make the oil get hot which means that the density gets less, which reduces the force (damper 'fade'). Sometimes dampers have extra reservoirs, to keep the fluid at a reasonable temperature. It does not help to have it mounted in a stream of hot air coming off an exhaust, radiator or the engine itself. All dampers fade to some degree. In tests made for *Performance Bikes* on a dynamometer (which drives the damper through a given stroke at a constant speed and measures the force) we found that one type faded noticeably after two minutes, while others, although they lost force, still kept an acceptable level of damping.

If the oil is drawn back too quickly it may cavitate and may also draw air bubbles through with it. This lowers the effective viscosity of the mixed fluid and also makes the fluid compressible, which does little to promote the efficiency of the unit. To avoid this problem, manufacturers have tried keeping the fluid in a pressurized reservoir and have also put a floating piston (de Carbon) or a membrane between the gas and the liquid. A heavier oil is more likely to cavitate. In general, it is better to use a lighter oil (SAE 5 W) and to increase damping forces by using smaller orifices.

Cartridge dampers – in which the damper piston and valves are inside a chamber – first appeared on long-travel motocross forks and are now fitted to all high quality forks. The advantage is that the damper mechanism is kept away from the surface of the oil so that aeration is not a problem and so that the damper piston can build up positive pressure in both directions, reducing the likelihood of cavitation during the extension stroke.

In rear shocks these problems are avoided by having the fluid pressurized, usually with a remote cylinder containing nitrogen gas at high pressure, a valve to maintain pressure and a diaphragm to separate the gas and the liquid. The pressure is added to the spring force but it only acts across the relatively small section of the damper rod – in telescopic forks it would act on the full cross-section area of the fork legs and would produce large spring loads. In cartridge forks, as in rear shocks, the damper consists of a piston assembly carried by the damper rod in a cylinder, sealed by a piston ring. When the forks are over-hauled, and refilled with oil, the cylinder chamber usually needs to be primed by slowly moving the damper rod up and down while keeping the oil topped up to the level of the chamber.

When the piston moves down (compression or bump stroke) oil passes through passageways in it and also in a fixed block at the bottom of the chamber. There are relatively large passageways which cope with the oil flow at low piston speeds. Where bump damping adjusters are provided, they are usually screws with tapered ends which project into one of these passageways. Consequently bump damping adjusters are usually found at the lower end of fork legs and the top (or in the remote reservoir) of rear damper bodies, and turning the screw in, i.e. clockwise, will increase the low-speed damping force. An alternative type of adjuster takes the form of a cylinder with four or more holes of different sizes; as it is turned, the holes line up with the low-speed passageway, creating a greater or smaller orifice for the oil to flow through. Both types of adjuster usually have spring-backed ball bearings which press against flats on the adjuster or the damper leg to give click-stop adjustment, but if the second type is turned one click beyond the maximum setting, it will go to the minimum setting and vice versa.

As the piston speed increases, the low-speed passageways reach full capacity and pressure builds up in the high-speed passages inside the piston. These are blocked by sprung shims, usually in a stack of various diameters; the fluid flow will lift a large diameter shim more easily than a stiffer, small diameter one, so the stack arrangement can be tailored to give whatever progression of lift/flow is deemed suitable.

Other than stripping the damper and changing the shim stack, high-speed bump damping is not usually adjustable. One or two high quality dampers do provide adjustment, usually in the form of a threaded rod which is tightened down on to the shim stack, increasing the load on it. However, the low-speed adjustment will affect high-speed damping by bringing the shim stack into play at a higher or lower piston speed and there is usually the option of using a more or less viscous oil.

Rebound (or extension) damping is also controlled by passages in the piston – but not the same ones. The shims are usually set so that they bear down across the open end of the orifice and make a one-way valve. Fluid coming through the passageway can lift the shim, against the cumulative spring force. But when the damper tries to push fluid back in the reverse direction it merely pushes the flat shim into harder contact with the piston.

Consequently the rebound shim stack is to be found on the opposite side of the piston. It works in exactly the same way and, if adjustment is provided, it will also be by a threaded adjuster which is tightened down on to the stack to increase the spring force, and therefore the damping force. Low-speed rebound damping is handled by large passages in the piston, and will have either a tapered needle or multi-orifice type of adjustment, as with the bump adjuster. Note that in many designs the low-speed passageways are not closed off by one-way shims, or are only partially closed off. In this case some fluid will pass through them in both bump and rebound directions and increasing the low-speed rebound adjustment will also make a (smaller) increase to bump damping.

On many cheaper dampers, rebound damping is the only adjustment provided, but this will often change bump damping as well.

The amount of fluid inside the damper has some bearing on the damping force. The less fluid there is, the hotter it will run and the damper will be more prone to fade. There must also be enough fluid to lubricate the mechanism properly, including the seals.

The level of fluid inside a sealed unit also affects the compression ratio – if the level is too high the pressure may get so high that it effectively locks the unit and prevents any more travel. Some fork oil is designed to react mildly with the material used to make the seals, causing it to swell slightly and therefore seal more efficiently. If the seals dry out they can contract, and running dry will obviously increase wear at the seal.

The intention behind the damper is to take some of the force that would have been put into or taken out of the spring. This force is transmitted directly to the other end of the unit, but its energy is absorbed and frittered away. It slows down the motion, helping to reduce jolts and attitude changes by spreading them over a longer period of time. This is all very well, except that while the damper is still dissipating the energy from one shock, another may come along, and then another. Before the spring can extend to its normal position, it will be compressed. The unit will gradually shorten – called 'pumping down' – and so lower the bike's ride height and use up suspension travel. This is caused by having too much extension damping.

It is possible to design the damping circuit as a pump and to use this to pressurize the shock absorber. Boge do this with their Nivomat system, which 'pumps up' until the ride height reaches a certain level, irrespective of the load carried by the vehicle.

Damping can be adjusted in several ways:

1   By external adjusters, if provided. Most units have little or no compression damping and the adjusters for this rarely make any perceptible difference. Koni struts are a notable exception. Extension damping is much greater, and adjusters on this stroke do make a difference.
2   By varying the viscosity of the oil. A heavier grade will increase the damping in both directions but is more prone to aeration and cavitation.
3   By increasing the amount of oil. In some designs this can reduce the likelihood of aeration or fade because it alters the spring rate and internal pressure. But the effects on springing are likely to be more noticeable.
4   By rebuilding the damper with different orifices, different spring shims and/or different tapered needle settings. This is a specialist's job, particularly where pressurized shocks are concerned.
5   By changing the spring rate. This will alter the deflection of the spring for a given load and therefore alter the speed at which the unit compresses. A harder spring will give less movement and therefore less damping.

Shock absorbers wear out because the fluid deteriorates, sprung shims may take a set, bottoming may damage part of the mechanism and seals and piston rings wear. Many of the competition types are rebuildable and should be overhauled whenever their performance begins to fall. Most specialists who can rebuild these units will have a dynamometer which can measure damping force and compare it to the as-new condition. Some specialists will rebuild the shocks with tighter clearances and their own seals, as well as re-valving them to alter the damping characteristics.

72

Linkages and levers

Springs of all persuasions act essentially in tension and compression; bending and shear forces have to be contained by the suspension legs, with, for example the bushes in the sliding legs of telescopic forks. The uni-directional spring force has to be provided by some kind of linkage and this gives the opportunity to alter the size and the travel of the sprung movement.

First, the linkage has to be strong enough to contain thrust and brake forces. Second, it can be used to amplify wheel travel/spring force and, third, it can be used to vary the spring leverage as the suspension deflects – altering the wheel rate or making the suspension load-sensitive.

There are several advantages to be gained here. First, the spring and damper unit can be mounted so that it has minimal effect on the unsprung inertia; the original swing arm design with one damper on the end of each arm was about as bad as it could be. Using the swing arm as a lever to operate a spring in front of the rear wheel meant that only one unit had to be used, and its travel, hence inertia, was one-half to one-fifth of the travel of the original two units. Second, the mass of the spring/damper can be placed closer to both the centre of gravity of the bike and to its roll axis. Third, the amount of damper travel can be regulated to give an optimum range of damping adjustment. If a damper was fixed to the end of a swing arm with 12 inches of wheel travel, the displacement inside the damper would be enormous and it would be difficult to build one which did not overheat or aerate its fluid. Fourth, the use of bell-cranks to operate the spring provides a much wider range of adjustment for pre-load, ride height and the rising rate effect. (Because you can use different bell cranks, adjustable tie-rods, mount

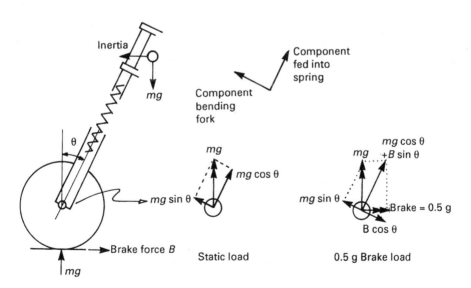

Figure 5.13 The forces in a single telescopic fork leg, carrying a load *mg* and brake force *B*. Components of the forces in line with the spring tend to compress it. Components at 90° to the spring merely tend to bend the fork leg

the cranks on adjustable eccentrics and so on, in addition to the normal spring and damping adjustment at the shock absorber itself).

A comparison of telescopic forks and swing arm suspension shows how the loads can vary, even in apparently similar systems. Consider a single telescopic leg as in Figure 5.13, inclined at an angle θ to the vertical and carrying an axle load of W (= mg). Compare this to a single-sided swing arm, with the spring unit mounted at the spindle, also inclined at an angle θ to the vertical and carrying a load W.

In the telescopic leg, the force required in the spring is $W \cos θ$, while the leg is subject to a bending force of $W \sin θ$ (multiplied by the length of the leg). During braking, a brake force B is partly fed into the suspension ($B \sin θ$) and partly used to bend the leg ($B \cos θ$ in the other direction to the bending moment caused by the axle weight). So, contrary to popular belief, there can be less bending moment in the fork legs during braking than when the bike is travelling at a steady speed. The required spring force gets larger as θ gets smaller, and is always smaller than the axle load, W.

A swing arm – for front or rear wheel – contains totally different forces, even though the geometry appears to be the same (see Figure 5.14(a)). Here the spring strut is carried on zero-friction bushes (in theory) and can only take axial forces. No bending moment can be applied to it. The wheel is located instead by the arm, which acts as a radius arm. Here, the required spring force is $W/\cos θ$, which reaches a minimum when θ = 0 and is, otherwise, always larger than W.

Thrust forces (drive for a rear wheel, braking for a front wheel) are all contained in the swing arm (well, nearly all; see anti-squat, below).

The greater the angle θ, the stronger the spring has to be (or the greater the leverage the wheel has over it) which is in direct contrast to the telescopic leg where a greater angle simply means that more axle load is devoted to bending the leg and less is put into the spring. As the telescopic fork compresses, the bike pivots about the rear wheel and the angle θ decreases – by roughly 1° for every inch of compression. This means that a stronger spring would be needed and actually creates a falling rate, albeit a very small one (and one which is totally offset by the rising rate caused by air pressure inside the forks in real life).

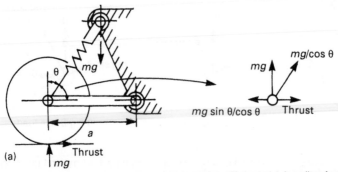

Figure 5.14 (a) The forces in a swing arm suspension. There is no bending force on the spring strut but the force fed into it is greater than the load *mg* and becomes greater as the angle θ

74

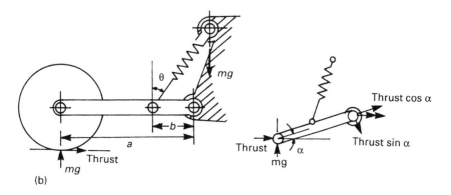

(b)

Figure 5.14 (a) continued

increases (which it does as the suspension compresses). Thrust forces are taken by the swing arm **(b)** If the spring is mounted away from the wheel spindle, the same things happen but now the spring force is increased by the ratio a/b (and its movement is reduced by the same factor). The spring force also tries to bend the swing arm. If the swing arm is not horizontal then there will be a vertical component of the thrust force produced at the swing arm pivot, which can either work with or against the spring force, depending on the direction of the thrust (accelerating or braking) and the angle of the arm above or below the horizontal. **(c)** Rear suspension linkage. This 1934 Matchless V4 used a triangulated swing arm to compress spring units – not unlike the mechanism employed by Yamaha some 40 years later.

Figure 5.14 (c)

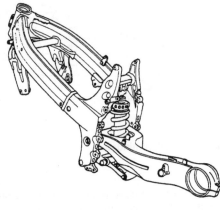

Figure 5.14 (d)

Figure 5.14 (e)

(d) Suspension linkages became more refined. The Suzuki RG500 used a rocker arm at the top and a bottom link to compress both ends of the spring. Honda evolved single-sided swing arms for their RVF racers, then the production VFR750R (RC30) and for this VFR400R

(e) Various attempts to beat the unsprung mass of telescopic forks, their rigidity and/or to transmit thrust without bending. Eric Offenstadt's TZ350–powered racers used trailing link forks (1976–78). (f) various Elf designs between 1978 and 1988 used a swinging arm arrangement. This is an early, RCB–powered endurance racer. (g) the Bimota Tesi prototype of the early 1980s finally went into production about ten years later

Figure 5.14 (f)

Figure 5.14 (g)

When the swing arm strut is compressed, the angle θ gets smaller and so does the required spring force. The wheel has worse leverage over the spring than before and this produces a natural rising rate. Also the pivoting arm is in the region of 16 inches long, so every inch of wheel travel produces an angular change of 3° to 4° at the swing arm (and at the strut).

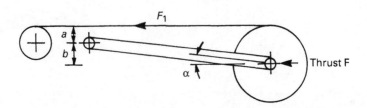

Figure 5.15 Anti-squat torque. The angle of the swing arm and the chainline create a torque ($Fb - F_1a$) which tends to extend the suspension

There is also scope to increase the leverage (and to reduce the spring travel, to give less inertia at the shock absorber) by mounting the strut part way along the swing arm - at distance b from the pivot, compared to distance a for the length of the arm.

Now the required spring force is $Wa/b \cos \theta$ and if the strut is mounted half way along the arm (so $a/b = 2$) then double the spring force is needed. Note that this would need a spring *rate* four times as high as before, because the leverage is doubled and the distance moved by the spring is halved. If any kind of leverage is altered so that the required spring force is n times as high as the original, then the required spring rate is n^2 times the initial rate and the distance moved by the spring is $1/n$th the travel of the original.

The same principle can be applied to the Monoshock type of linkage used by Yamaha (and Matchless and Vincent, etc) and to the more complicated linkages, except that as it gets increasingly difficult to calculate the leverage it becomes easier to measure it from a drawing.

When the leverage increases, the spring travel descreases and therefore the spring/damper velocity decreases as well. As damping force is proportional to velocity squared, the new damper has to have considerably more damping capacity. If a longer – or shorter – swing arm is fitted it can change either the leverage or the fitted length of the strut, or both. The required spring force is that needed to support the weight on that axle. The height at which the bike sits is its ride height (usually simplified to some convenient measurement between a point on the sprung mass and another point on the unsprung mass – see Chapter 2). The force which supports it comes from the spring's pre-load and, once this has been exceeded, from the spring rate.

Pre-load, the fitted load in the spring, is the force which must be used to make the spring move at all. If there is too much pre-load, the weight of bike and rider will not compress the suspension far enough (it will have no reserve extension travel) and, ultimately it will ride topped out. In some

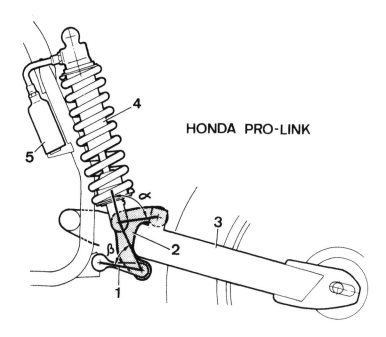

HONDA PRO-LINK

Figure 5.16 (a)

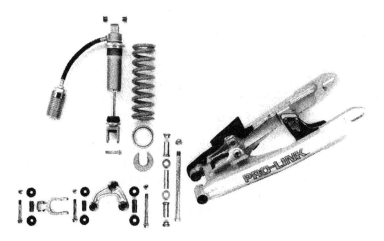

Figure 5.16 (b)

Figure 5.16 (a) and (b) Evolution of rising rate suspension: Honda used a radius arm (1) and a bottom link (2) to control the leverage of the swing arm (3) over the spring strut (4). The reservoir (5) contains gas at high pressure behind a diaphragm, in order to pressurize the fluid, reducing its tendency to aerate or cavitate during rapid damper movement. (*Honda*)

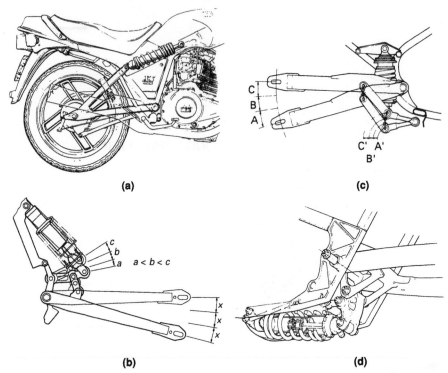

Figure 5.17 Yamaha initially used a simple yet very effective linkage (a) whose drawback was that it took up a lot of room. It was followed by radius arm linkages (b) on their off-road (TY250) and (c) road bikes (RD350LC) to place the spring strut vertically, behind the engine and, (d) on the RD500LC, horizontally, mounted below the engine. (*Yamaha*)

conditions – motocross or enduro on fast, bumpy circuits – bikes feel better this way because it leaves the maximum compression travel to handle bumps and, when the machine jumps, the suspension tops out anyway.

Pre-load adjustment is also a convenient way to control the ride height which affects both ground clearance and steering geometry. On a typical wheelbase of 55 inches, a change of 1 inch in ride height at one wheel or the other will move the centre of gravity by about half an inch and will alter the steering castor by about 1° (as the normally-used range is 24° to 30°, this is a significant amount). Making the castor steeper in this way will also reduce the trail.

The spring rate is the force needed to compress the spring by one unit of length; the wheel rate is the force needed to raise the wheel one unit of length against the spring force. If the leverage ratio is n (measured from the bike or taken from a drawing) then, assuming one spring:

wheel travel	$= n \times$ spring travel
force in spring	$= n \times$ axle load
spring rate	$= n^2 \times$ wheel rate
damper speed	$=$ (vertical) axle speed$/n$

80

If the load is shared between two springs in parallel then the spring force and spring rate will be halved. The implications of the damper speed are that its inertia will be reduced by a factor of n and the damping force will be reduced by a factor of n^2. This applies to any linkage but in some linkages the factor n may change as the suspension is deflected.

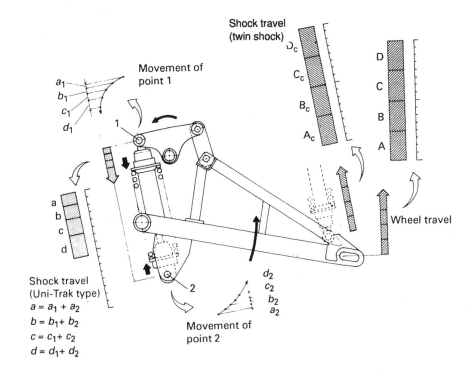

Figure 5.18 Kawasaki used a rocker arm mechanism on their early Uni-Trak. (*Kawasaki*)

If the axle load and the spring pre-load are known, the static position of the suspension can be calculated from the spring rate and the leverage factor, n.

Anti-dive suspension

Weight transfer under braking makes the front suspension compress, changing the attitude of the bike and, if the motion is fast enough, making it difficult to control. Anti-dive can achieve two aims here: to stiffen the compression damping during braking to slow down the dive; and to regulate the height of the centre of gravity during braking in order to get maximum traction. Both of these aspects are covered in detail in Chapter 6.

1   If the bike is able to lift its rear wheel under braking, then more dive would be beneficial.

2   If the bike is able to lock its front wheel or if its braking performance is significantly better when the rear brake is used as well as the front, then less dive would be beneficial.

It is quite likely that a machine will meet condition 1 when the track is dry, and condition 2 in the wet.

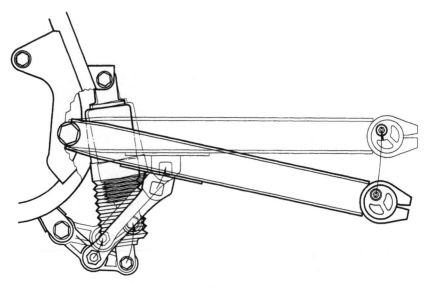

Figure 5.19 Other Kawasaki suspension systems reverted to the bottom link/radius arm design. (*Kawasaki*)

Anti-squat suspension

During acceleration, the full weight of the machine can be transferred to the rear wheel. The suspension needs to be able to carry this load *and* maintain the centre of gravity at a height which will give maximum traction. If the centre of gravity is too low, the wheel will spin too easily; if it is too high, the bike will overturn too easily.

If the force between the tyre and the road is F, this reaches a maximum of $F_{max} = W_r\mu$, where W_r is the rear axle load and μ is the coefficient of friction between tyre and road. If F exceeds this value, the tyre will spin. If the engine horsepower *at this speed* is hp, and the road speed is v mph, then the force (lbf) at the tyre is $F = 375$ hp/v.

Up to the point of wheelspin, F is the thrust which makes the bike accelerate and will set up an overturning couple, Fy, where y is the height of the centre of gravity, which makes the bike try to wheelie. It is opposed by the weight of the bike, mg which produces a maximum couple mgx about the back wheel, where x is the distance of the centre of gravity in front of the tyre's contact patch. This value is reached when all weight is taken off the front wheel (in which case $W_r = mg$).

82

If $Fy > mgx$ then the bike will wheelie
If $F > F_{max}$ then the wheel will spin

The weight transfer on to the back axle is given by W_r:

$$W_r = mg + Fy/w - mgx/w$$

where w is the wheelbase.

Therefore $W_r = mg$ when $Fy = mgx$.

From the axle load, the force on the rear spring can be calculated, if the lever ratio is n (see above):

Force on spring $\quad = W_r n$

Spring deflection $\quad = (P - W_r n)/s \quad$ (total)

$\qquad\qquad\qquad\quad = (W_r - W_{ro})n/s \quad$ (change from static spring length)

Where P is the pre-load, s is the spring rate and W_{ro} is the static rear axle load.

This is only a first approximation as the spring compression will lower the centre of gravity slightly and this will have an effect on the calculations although front spring extension will raise the centre of gravity. There is also a second-order torque which increases the overturn or wheelie torque and is not allowed for in these calculations. It is produced by the reaction to making the rear wheel accelerate – the tendency for the wheel to stay still and for the bike to rotate around it. It is proportional to the inertia of the wheel in relation to the whole bike, and to the rate of acceleration. It would tend to raise the nose of the bike if the throttle were opened while the bike was jumping and not in contact with the ground. There is another force which may alter the suspension travel and this can be used deliberately to prevent squat during acceleration. It is caused by the swing arm not being parallel to the ground, while the thrust between tyre and road is horizontal. The wheel bearings can only transmit a horizontal force to the spindle and the swing arm, but if the swing arm is not level then the thrust in it will produce a couple Fb, where b is the height of the swing arm pivot above the wheel spindle, see Figure 5.15.

If the pivot is higher than the spindle, the couple Fb will try to lift the bike and extend the rear suspension. At the same time, the pull in the chain line creates a similar, but opposite couple, $F_1 a$, where F_1 is the force in the chain and a is the distance between it and the swing arm pivot. In essence, the swing arm and drive sprockets form an over-centre mechanism. When all three centres are in line, the chain is at the position of maximum tension. If

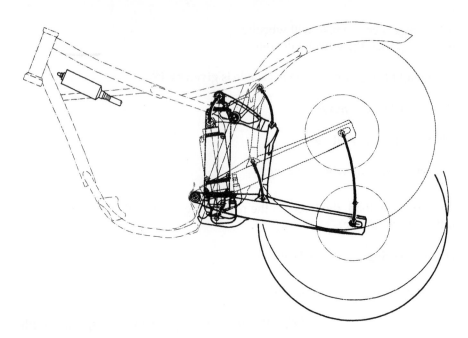

Figure 5.20 Suzuki's Full Floater used a bottom link and a rocker arm at the top, to compress the spring from both ends. (*Suzuki*)

the wheel is displaced to either side, the pull of the chain will tend to displace it further, see Figure 5.15.

These forces set up a torque in the swing arm:

Anti-squat torque $= Fb - F_1a$

and $F_1 = Fd_2/d_1$

Where $d_{1,2}$ are the diameters of the wheel sprocket and the rear tyre, respectively. A torque in the swing arm divided by the length of the arm (L) is equivalent to a force acting at the spindle, and this multiplied by the lever ratio, n, is the equivalent force at the spring unit.

Anti-squat force at spring $= (Fb - F_1a)n/L$

So the total force at the spring is $W_rn - (Fb - F_1a)n/L$

This works without chain drive i.e., $F_1a = 0$ as in the case of pre-Paralever BMW shaft drive twins which generate large anti-squat forces during acceleration. The Paralever mechanism takes some of the torque needed to restrain the bevel gear housing and transmits it directly to the sprung part of the chassis, in a way which tends to compress the suspension during acceleration.

Figure 5.21 Racers made fabricated links and rocker arms in order to adjust the ride height and spring leverage

The spring force can now be used to work out the spring movement under acceleration and the height of the centre of gravity. The spring rate, the spring pre-load and the anti-squat torque can be tailored to produce the height of the centre of gravity which gives maximum traction. A spreadsheet program (similar to the one in the Appendix which does a calculation for maximum braking effort) is probably the easiest way to handle this.

Maximum traction occurs when:

$F = mgx/y$ (point of wheelie), and

$F = mg\mu$ (point of wheelspin, all weight on rear axle). That is:

$x/y = \mu$

So in conditions where $\mu = 1$ (race or sports tyres on a good surface), the height of the centre of gravity needs to be equal to its distance from the rear wheel spindle. When $\mu = 0.5$ (wet conditions for example) then the centre of gravity needs to be moved back or raised until its height is twice its horizontal distance from the rear spindle. If tests show a tendency to wheelspin, then the centre of gravity should be raised or moved back and/or more anti-squat built in. If the bike is prone to wheelie under acceleration, then the centre of gravity should be lowered, shifted forward and/or less anti-squat used.

Figure 5.22 Inverted forks – with the lighter stanchions at the bottom and the heavier sliders and damping mechanism carried at the top – appeared in 1987–88. The thicker, stiffer section of the slider is now mounted in the yokes, where the bending moments are greatest, improving fork stiffness and unsprung weight. Carbon fibre discs and composite wheels made the unsprung mass smaller still

The requirements for best tractive effort may not be compatible with those for optimum braking (see Chapter 6). In this case it is necessary to decide which is the more important, to compromise, or to build the machine so that the rider can move far enough forwards and backwards to shift the centre of gravity as required. A change in wheelbase may allow both requirements to be met.

Telescopic forks have several weaknesses:

1   Steering, braking/engine thrust force and suspension loads are not separated, and each force is passed through the other components, meaning that they have to be unnecessarily strong (heavy) in order to accommodate all the forces.
2   Forks permit long suspension travel but the length of this travel must be added to the height of the fork clamp (or yoke) above the tyre. This makes the forks long, giving brake forces, etc, a large bending moment about the fork clamp and the steering head. Both the forks and the steering head must be made stiff enough to resist this, which means they tend to be heavier than

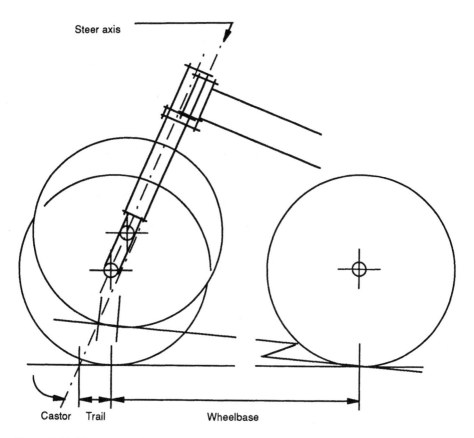

Steer axis

Castor   Trail   Wheelbase

Figure 5.23 When conventional telescopic forks compress, the wheelbase and trail are reduced and the castor angle becomes steeper. This effect is increased if the rear suspension extends (e.g. under braking) but is reduced if the rear suspension compresses (e.g. while cornering)

they might have been. Also the frame has to go from the swing arm pivot up to the steering head, a longer route – requiring more material – than the shortest, which would be a straight line drawn between the wheel spindles.

3 Because of all this, the unsprung mass and the steered inertia are both greater than they might have been.

4 There is virtually no control over steering geometry changes when the suspension compresses (although the natural tendency of telescopic forks to make the castor steeper and reduce the trail as the front axle load increases, seems to be the right thing to do).

Alternative designs usually have two aims:

1 To separate the forces in the suspension, braking and steering.

2 To create a more direct load path and so, for the same stiffness or strength, to produce a lighter frame and lighter unsprung components (or produce stiffer components for the same weight).

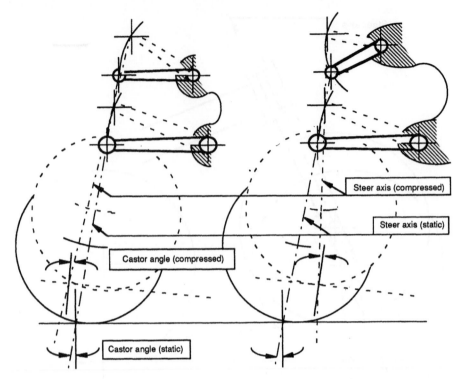

Steer axis (compressed)

Steer axis (static)

Castor angle (compressed)

Castor angle (static)

Figure 5.24 Suspension controlled by wishbones or pivoted links gives a lot of control over steering geometry. Depending on the lengths of the links and their angles, the castor and trail can be reduced, increased or kept constant as the suspension deflects

They tend to fall into two categories: those which use a wishbone to control the front suspension, pivoting just above the tyre with an approximately horizontal link, and those which have a swing arm, not unlike rear suspension, with the steering pivots built inside the diameter of the wheel (hub centre steering).

The latter type has the most direct load path and thus requires the simplest (lightest/stiffest) frame, typified by designs such as the various Elf racers, Bimota Tesi, Yamaha GTS. The lower steering joint is carried on the swing arm, the upper one on a parallelogram linkage, the proportions of which give the designer scope to make virtually any geometry change with suspension movement. Wishbone suspension has a less direct load path but has advantages in giving a better steering lock than hub centre steering without excessive chassis width.

Both types have advantages in unsprung mass and steered inertia, both allow the geometry to be varied at will – by altering the relative lengths and angles of the wishbones/parallelogram linkages – and both can operate single damper units via bell-cranks or rocker arms (similar to conventional rear swing arm linkages) which can give rising rate suspension and doesn't subject the damper to full wheel travel, as telescopic forks do.

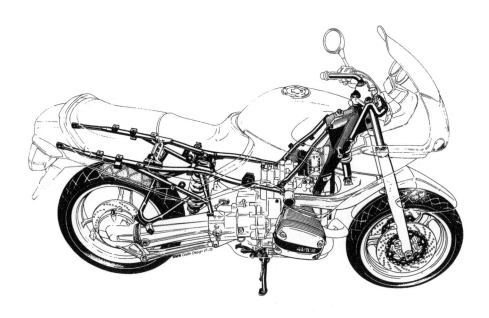

Figure 5.25 Single wishbone and slider fork design used by BMW on the R1100RS. Castor and trail are kept approximately constant when the front suspension compresses. *(BMW)*

89

In addition it is easy to incorporate anti-dive mechanisms (see Chapter 6) either by feeding brake torque into the suspension or by separating the forward inertia force from the load carried by the front suspension by selecting the angle and position of the lower wishbone or swing arm.

Both types suffer the disadvantage that it is difficult to get more than about five inches of wheel travel without using very long swing arms or wishbones.

Figure 5.26 Hub centre steering as used by Yamaha on their GTS. While forming a very rigid frame and suspension assembly, it does not take advantage of this to save weight. Changes in steering geometry emulate telescopic forks. *(Yamaha)*

Chapter 6

Brakes

Brakes are among the simplest things on the bike yet, in some respects, they are the most powerful part of it. If a bike can accelerate to 120 mph in a quarter of a mile, it can stop in less than half that distance. Tests show that most bikes can decelerate at more than 1g, which means that the brakes have to generate a force greater than the bike's weight. This is restricted by the grip at the tyres, of course, and the brakes are often capable of more.

The brakes convert kinetic energy into something else – heat – and a look at kinetic energy levels will show what they are up against. It is defined as one half of the total mass, multiplied by the square of its velocity, $0.5\ mv^2$. Consider a GL1500 Gold Wing at its gross train weight of 1270 lb, and an RS250 at its possible race weight of 340 lb. Table 6.1 shows what is required of the brakes.

Table 6.1 Kinetic energy comparison

Bike	Weight lbf	Mass m lbm	Speed mph	v ft/s	v^2	$\frac{1}{2}mv^2$ ft lbf
GL1500	1270	39.4	55	81	6561	129,251
RS250	340	10.6	106	155	24,025	126,840
GL1500			106	155	24,025	473,292
RS250			204	299	89,401	473,825

The final column represents kinetic energy and its units – mass × ft²/s² – could be written as mass × ft/s² x ft. Mass × acceleration is force, so it becomes ft lbf, which is the same unit as torque but of a slightly higher order than the 70 or 80 ft lbf produced by the Gold Wing's engine.

Let us start with the Gold Wing trundling along at 55 mph, the US national speed limit. That is 81 ft/s. Squaring it gives 6561 and multiplying by half the mass (mass is weight divided by g, gravitational acceleration), gives the final energy figure, 129,251 ft lbf. (The inertia of the rotating parts should also be included with the mass of the machine; it has the effect of making the mass appear to be greater, in proportion to the speed). To get roughly the same energy level, the RS250 has to travel at 106 mph. In other words, the Gold Wing will take as much effort to stop from 55 mph as the RS250 will from 106 mph.

If the Gold Wing speeds up to 106 mph it has, predictably, about four times the energy of the RS at the same speed. If they could stop from this speed in 5 seconds (which is 155/5 or 31 ft/s^2, nearly 1 g and it is about half the time it would take to accelerate to that speed, so it is a reasonable assumption) then we can estimate the power developed at the brakes. Power is a *rate* of doing work. One horsepower is defined as 550 ft lbf/s, so if the brakes converted 126,840 ft lbf in 5 seconds, that would be equivalent to 25,368 ft lbf/s or 46.1 horsepower, for the RS250's brakes. Not so hot, really, as that represents the total effort and some of it would come from air drag and rolling resistance. But the same calculation for the Gold Wing shows that it needs to dissipate 172 horsepower to stop at the same rate.

The braking force is applied at the tyre, but it is generated at the disc or drum, and the wheel acts like a lever. If the drum or disc has half the radius of the wheel, then only half the disc force will be available at the tyre. Alternatively, for every pound of force produced at the tyre, two pounds have to be made at the brake. Now if the disc or drum could be given the same diameter as the tyre, this ratio would become 1:1 instead of 2:1 – you can see the attraction of small wheels and large discs. Also, with the necessarily large forces produced at the disc, a high degree of feel and progression is essential for the rider to be able to control it.

There are many variables which affect this rather delicate balance of brute force and control, as Table 6.2 shows.

Table 6.2 Brake components

Part	Effect on brake	Other effects
Wheel diameter and rotor diameter	Torque, rubbing speed	Unsprung weight, inertia
Rotor material	Friction, wear, heat dissipation	Unsprung weight, inertia
Pad material	Friction, wear, heat dissipation	Unsprung weight
Pad size	Local pressure, wear	—
Caliper	Force at pad, hydraulic ratio	Unsprung weight, steering inertia
Master cylinder	Hydraulic ratio (hand force/brake force)	Steering inertia, prone to crash damage
Brake line(s)	Deflection under pressure, 'feel'	Suspension, steering travel

In general, the force applied at the lever or pedal is used to pressurize fluid (or pull a cable, etc.) and the resulting pressure is used to push the brake friction material against the moving rotor, whether it is a disc or a drum. There is a certain amount of deflection or springiness, which we discern as 'feel'. If there is too much or too little, the brake becomes difficult to control and feels either spongy or solid.

There is also weight transfer. The brake force opposes the motion of the bike, the inertia of the bike (and rider) want to keep travelling at the same speed. As the bike is slowing down, the rider feels as if it is moving backwards against him or as if he is being forced towards the front of the bike. The bike's inertia acts in a similar manner; it acts through the centre of gravity, which is probably 25 to 30 inches above ground level while the brake force appears at the tyres, at ground level. The result is an overturning torque (or couple) which tries to rotate the bike forward on to its nose. The bigger the force and the higher the centre of gravity, the greater this over-turning tendency will be.

The first effects are that the rider's weight is thrown forward on to the handlebar and the bike's weight is also thrown forward, compressing the suspension ('dive'). The rider can feel the force and sense the dive; if this varies in proportion with the brake lever pressure, then the brake feels progressive. If there is also a discernible movement at the lever when he squeezes harder, yet the resistance increases against his pressure, then the brake is giving a predictable feel, the rider can tell what is happening and – more importantly – what is going to happen next.

The second effect of all this is that weight is shifted to the front wheel and therefore removed from the back wheel. Traction at the front increases, while that at the back decreases. If the front brake pressure is steadily increased, the deceleration increases, and so does the weight transfer. One of two things can then happen as a limiting condition; either the front wheel will lock and skid, or it will continue to grip and the bike will stand on its nose, the back wheel leaving the floor. It depends on the amount of grip between tyre and road, the height of the centre of gravity and the weight of the bike. There is a computer spreadsheet program listed in the Appendix which will run through the various combinations, predict lock-up or overturn and suggest optimum geometry for the highest usable braking force.

To see how the system operates in detail and how it can be improved, it is necessary to go through it piece by piece, in sequence, starting at the lever and following it through to the brake pads and the tyre, as shown in Figure 6.2.

1 Lever

The force P is exerted by the rider's hand, or foot in the case of a pedal. The exact position from the pivot (x) varies with his hand position and how many fingers he uses. The force F_1 is fed into the master cylinder or cable so that:

$$F_1 = Px/y \tag{6.1}$$

F may be increased by increasing P – pulling harder (or positioning the lever so that it is easier to exert more force on it), by increasing x or decreasing y. Note that the dimensions x, y are at right angles to the lines of force, and change as the lever turns on its pivot. Giving the rider more leverage is an easy way to increase overall braking force but it does create more travel – if he positions his hand further along the lever, for example.

Figure 6.1 Overturn torque increases the load on the front wheel and reduces it at the rear

The lever can be curved (dog-leg lever) to give a more comfortable grip (or to clear the twist-grip) and there is often a span adjuster. It should be set so that the rider can move his fingers to the brake quickly and so that they form a natural angle with the hand and wrist. He will still need to be able to control the throttle, even when exerting full pressure on the brake.

There is no need for the lever to be any longer than the last finger position; it is a good idea to move the lever inboard as far as possible to prevent damage in a crash or in a collision with another rider (which could apply the brake, or damage the lever). The handlebar clamp should allow the lever to twist on the handlebar rather than bend or break. As the clamping screw will not be fully tightened, it should be locked.

It is also important that the lever, its pivot and the master cylinder/cable arrangement is completely rigid. Any flexure here will show up as lost motion and sponginess at the lever.

Other problems are mainly to do with the lever coming right back to the handlebar under full braking, assuming this is not caused by air in the hyd-

94

raulic line or vapour owing to fluid boiling. BMW K100s suffer from this problem, where the brakes are perfectly good for normal braking, but have too much travel for full braking. One cause can be lack of rigidity in the flexible hoses (or in a cable and linkage) in which case stronger items are needed (see brake lines, below). Hydraulic brakes are self-adjusting, while most cable-operated brakes are not, but there can still be too much clearance, caused by disc runout or by the rubber seals being too strong (see calipers, below). If adjustment fails to cure the problem, the final choice is to move the lever further from the handlebar, or to improve the braking force so that less lever force/travel is needed.

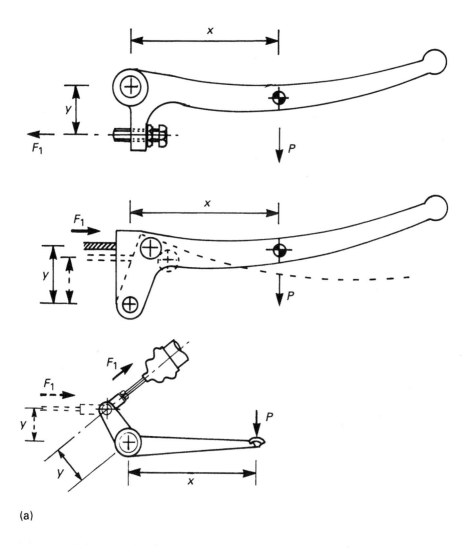

(a)

Figure 6.2 (a) Lever and pedal dimensions. This is the notation used in the text.

95

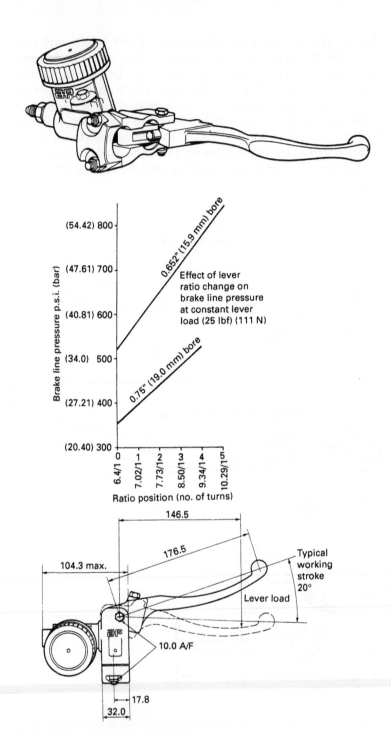

Figure 6.2 (b) AP Racing produced an adjustable lever which gave the rider a wide range of line pressure for a given hand force. (*Automotive Products*)

2 Master cylinder and reservoir

The significant dimension of the master cylinder is its diameter, d_1, which is sometimes stamped on the body of the cylinder. Its cross-section area A_1 is:

$$A_1 = \pi d_1{}^2/4$$

The pressure p created in the fluid by the force F_1 is:

$$p = F_1/A_1$$

This pressure acts on the piston of a slave cylinder to create another force F_2, so that:

$$F_2 = pA_2$$

or

$$F_2 = F_1 A_2/A_1 \tag{6.2}$$

where A_2 is the area of the slave piston. Note that this is the total area of all slave pistons. Where there is one piston, but the caliper slides on pins, bringing a fixed pad into contact with the far side of the disc, then the caliper is acting as a piston to operate the 'fixed' pad.

In this case $A_2 = \pi d_2{}^2/2$, if d_2 is the diameter of the piston. If the caliper has opposed pistons, then A_2 will be $\pi n d_2{}^2/4$, where n is the number of pistons. Note that in four-piston calipers, the leading pistons are often smaller in diameter but A_2 will be the total area of all of them.

Equation 6.2 shows how the force is magnified in proportion to the ratio between the area of the slave piston(s) and the area of the master piston. If A_2 is twice as big as A_1, then F_2 is two times F_1.

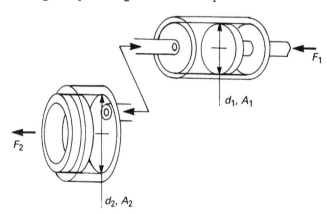

Figure 6.3 The hydraulic ratio is determined by the relative sizes of the master cylinder and the slave cylinder(s)

A_2/A_1 is the hydraulic ratio between the two pistons, and is the same as d_2^2/d_1^2, or $(d_2/d_1)^2$. It works exactly like a lever, and the original force P has already been magnified by a lever (see equation 6.1). Note that if the diameter of the slave piston is twice as much as the master piston, the hydraulic ratio becomes 4:1. If we put in the values from equation 6.1 we can see how the rider's hand force has been translated into the final force at the caliper, which will push the pads into contact with the disc.

$$F_2 = PxA_2/yA_1 \tag{6.3}$$

To increase the brake force F_2 it is necessary to increase P, x or A_2 or to reduce y or A_1. This takes no account of friction at the seals of the master cylinder and the slave cylinder(s), which will reduce F_1 and F_2 respectively. It will increase the required value of P or anti-knock-back springs can be used to compensate (see page 103). Note that if the pads have to travel a distance z before they touch the disc, the fluid displaced will be zA_2 and this volume will have to be moved at the master cylinder. The piston travel at the master cylinder will be z_1 so:

$$z_1 = zA_2/A_1$$

and the travel at the centre of pressure of the lever will be z_2 so that:

$$z_2 = z_1 x/y$$

or

$$z_2 = zA_2x/A_1y \tag{6.4}$$

So all the factors that increase the force F_2 will also increase the lever travel, z_2, in the same proportion.

The opposite case also applies. It is possible to reduce lever travel by making z (pad clearance), A_2 and x smaller or by making A_1 or y larger. All of these, with the exception of pad clearance, will also make the brake heavier in operation. ·

This allows for no flexing at the lever or cylinder, and no expansion of the hydraulic lines. Flexing adds directly to lever travel; expansion increases the volume of the hydraulic system, which has to be supplied by fluid from the master cylinder, thus adding to the required travel of the piston. Friction in the pins of sliding calipers or in a floating disc will be subtracted from the force available at the pad. It can be reduced by using a high temperature lubricant such as Copaslip.

The equivalent calculation for a drum brake gives a force F_2 at the brake shoe, where:

$$F_2 = Pxa/yb$$

and a/b is the ratio between the lever length and the cam height. As the lever turns, the ratio a/b can alter; the brake should be adjusted so that this is approaching a maximum when the shoe contacts the drum. If it is allowed to go over-centre, the ratio will start to decrease.

Early types of master cylinder were bulky, heavy and prone to damage. Later types are much more compact and many have either a remote reservoir or a built-in reservoir which can be used at a variety of angles before there is a risk of air getting into the cylinder itself.

The reason for having a reservoir at all is that fluid must be able to replace the volume left as the pads wear and air must be able to replace the fluid, otherwise the lever will not return to its neutral position, or if there is sufficient spring strength to force it, the pads will be pulled back too far from the disc, resulting in excessive lever travel as the pads wear down. A flexible, concertina-like seal is fitted between the fluid and the air above it, partly to prevent atmospheric moisture contaminating the fluid and partly to prevent fluid swilling out through the vent in the top of the reservoir.

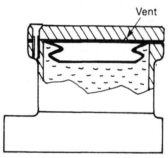

Figure 6.4 The diaphragm seal in the top of the master cylinder prevents air-fluid contact but allows the fluid to be displaced

Racing machines, which do not need to carry much fluid, often omit the reservoir and use a short length of flexible tube instead. This also needs an air vent and, because of its flexibility, needs a seal between the air and the fluid.

A small leak of fluid can be seen by what appears to be condensation on the outside of the reservoir. The fluid is hygroscopic – it attracts and absorbs water, and in doing so its volume increases. This can be seen by putting a few drops into a jar and watching it over the course of several days.

The slightest amount of dirt on the gasket face will allow the fluid to leak, while running the reservoir at too steep an angle is likely to allow air into the master cylinder, especially if vibration causes frothing.

3 Brake fluid

Most easily available brake fluids conform to standards SAE J1703, ISO4925, and FMVSS116 DOT3, DOT4 and DOT5. These standards set out minimum requirements for boiling points, viscosity, compatibility with seals, compatibility with fluids made to the same standard, and metal corrosion.

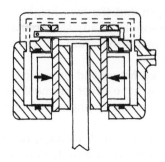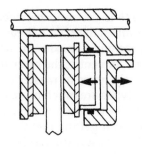

Figure 6.5 Principle of the two main types of caliper. *Left:* opposed piston, equal pressure delivered to cylinders either side of the disc. *Right:* floating caliper. Pressure in single cylinder forces piston towards the disc and pushes the caliper (which is free to slide on pins) away from the disc, bringing the 'fixed' pad into contact with the far side of the disc

The first three are broadly similar and are the most common recommendations for road machines (which is usually marked on the master cylinder/reservoir). The important feature apart from compatibility, is the wet boiling point, i.e. the boiling point when the fluid has been contaminated with water. Typically, 4 per cent water can lower the boiling point by 30 per cent. The SAE, ISO and DOT3 standards call for a minimum wet boiling point of 140°C while DOT4 requires 155°C. DOT5 was designed for silicone brake fluids which are not hygroscopic but which are said to be compressible at high temperatures and are therefore unsuitable for high performance use.

In addition to these fluids there are specialist fluids developed for competition, such as Ferodo Racing Brake Fluid and AP Racing 550 and 600 brake fluid.

AP's fluids conform to SAE J1703 but have higher boiling points: AP550 has a minimum 'dry' boiling point of 290°C and a wet boiling point of 145°C; AP600 goes from 300°C 'dry' to 210°C wet.

As a general rule, the higher the boiling point, the more hygroscopic the fluid – which means that it needs changing more frequently because the water contamination not only lowers the boiling point but can also cause corrosion in the cylinders. The manufacturers of racing fluids recommend that the fluid is bled thoroughly before each event.

To bleed the system, remove the caliper(s) so that the nipple can be held at the highest point of the entire system. Leave for a while to let air bubbles drift to the top. Air in the master cylinder can be expelled by moving the lever gently back and forth around the point where the port between cylinder and reservoir opens. Connect a clear pipe to each bleed nipple and run it into a jar containing brake fluid, so that the pipe cannot suck air back into the caliper. Put a piece of wood between the pads. If the design of the bleed nipple ensures that air cannot be drawn in past it, loosen the nipple slightly and steadily pump fluid through, keeping the reservoir topped up. If air can get back past the nipple, close it at the end of each stroke.

100

Air in the hydraulic system will be compressible and will use up a lot of movement at the lever, giving a spongy feel and ultimately restricting the pressure which can be reached. The line pressure can be increased by pumping the lever rapidly and the brake 'pumps up'. This will be a permanent feature until the air is removed by bleeding the system.

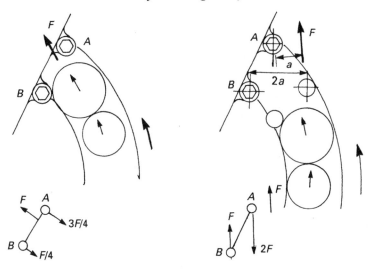

Figure 6.6 Relocating the caliper mounting points can seriously change the forces in the pins. In this case the force at the top pin is increased by 2.67 times while that in the lower pin is increased by a factor of 4 (for the same braking force). The direction of the force has also changed, putting the cast boss which holds pin *B* mainly in tension instead of compression (the material is much weaker in tension than it is in compression)

The same symptoms, but on a temporary basis, happen if the fluid is overheated and boils locally, forming a pocket of vapour. Lever travel increases, the brake feels spongy and cannot develop full pressure. The symptoms go away when the brake cools down and the vapour condenses. It is often called 'fade', but, while 'fluid fade' might be an apt description, it is totally different to pad fade (see pads, below).

4 Brake lines

Typical maximum line pressures are 500 to 600 lb/in^2 and this is enough to make flexible hoses expand. Some expansion is desirable: it creates 'feel' and it limits the build up of high pressure, making the brake more progressive. Too much expansion makes the brake feel spongy, creates excessive lever travel and prevents the brake from developing its full pressure.

The compromise may be exceeded if a roadster is used for racing, if performance is increased forcing the brakes to be used more heavily or if the brakes themselves are uprated. More rigid lines are needed and Goodridge UK is the main supplier in the UK, with a range of its own and Earl's braided hoses plus all the accessories to go with them.

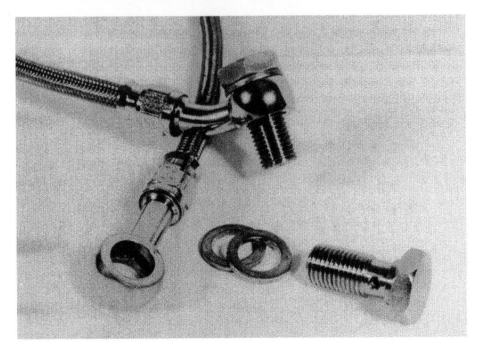

Figure 6.7 Various strengths of hose are available, with a choice of synthetic rubber, nitrile or PTFE inner hose backed by an outer braid of cotton, nylon, wire or stainless steel. (*Goodridge UK*)

For most applications it is important not to overdo it and make the feel of the brake completely solid. Some flexibility is necessary and it is usually possible to build up a combination of hoses which will give the required stiffness in expansion plus flexibility to follow suspension travel. The line should be checked carefully over the full suspension movement, and with full steering travel, to make sure that it cannot chafe, get trapped or get stretched. Goodridge will make up lines to a pattern, and supply cutting and joining instructions in their catalogue.

5 Calipers
There are several caliper designs:

(a) Single piston, floating caliper.
(b) Two piston, floating caliper.
(c) Single piston, floating disc.
(d) Opposed piston (two piston).
(e) Opposed piston (four piston).
(f) Four piston, unequal diameters.
(g) Six (or more) pistons.

The first three only have the piston(s) pushing against one pad, the caliper or disc slides to bring the other pad into contact. Of these, type (c) is quite rare now, while (a) is common on roadsters and low-performance

102

machines while (b) is used on roadsters, motocrossers, etc. These types have a distinct advantage in being lighter, cheaper and narrower than opposed piston calipers. Their disadvantage is that friction in the sliding mounting has to be subtracted from the braking effort, and this friction increases as the brake force increases. It is also subject to road dirt and corrosion. Frequent cleaning and lubrication with Copaslip is the only way to ensure floating calipers stay efficient.

Virtually all high performance machines have opposed piston calipers which are self-aligning, although in many cases the disc is mounted on dowels and has a limited freedom to float while it is also free to expand without distortion.

The piston seals are mainly responsible for retracting the pistons when the brake is released, although disc runout will also push the pads and pistons back.

Specialist suppliers have lubricants such as Lockheed Disc Brake Lubricant, a silicone compound which protects the pistons from corrosion and helps assembly without damaging the seals. If the seals pull the pistons back too far, causing too much lever travel, it is possible to fit springs behind the pistons (anti-knock-back springs). AP lists 0, 4 and 7 lb fitted load springs. It is worth checking first that there is no blockage where the master cylinder is fed from the reservoir because as the spring in the master cylinder pushes the lever back, the low pressure will tend to pull back the pistons if new fluid cannot enter the master cylinder.

Conversely, if the pads drag, then any springs should be removed or reduced, new seals fitted or a slight amount of runout introduced at the disc.

The pads are often shimmed with anti-chatter springs or with a heat shield; Nissin pads fitted to some Hondas have a ceramic shield plasma-sprayed on to the backplate. It may be necessary to reduce the heat path to the caliper and fluid if the performance of the bike and its brakes are increased.

If the coefficient of friction between the pad and the disc is μ_B then the braking force at the disc is given by $F_2\mu_B$, from equation 6.3. This sets up a brake torque in the wheel of $F_2\mu_B r_B$, where r_B is the effective radius of the disc. If r_w is the rolling radius of the wheel, then this is turned back into a force F between the tyre and the road surface, so that:

$$F = F_2\mu_B r_B/r_w$$

or, substituting F_2 from equation 6.3:

$$F = PxA_2\mu_B r_B/r_w y A_1 \tag{6.5}$$

This gives us the braking force at the tyre, in terms of the rider's force at the lever and the dimensions of the brake system. To increase the braking force at the tyre (F) it is necessary to:

Increase	Decrease
Lever force P	Tyre rolling radius r_w
Lever length x	Lever/cylinder length y
Caliper piston area A_2	Master cylinder area A_1
Effective disc radius r_B	
Pad friction μ_B	

This also shows why competition discs have evolved the way they have. First it was necessary to increase the radius relative to the wheel size. Then the four-piston caliper was developed because it allowed a smaller swept area to be used, giving a larger effective radius even if the outer radius of the disc was the same. See Figure 6.14. Piston area could be increased over a two-piston layout and yet the disc could be lighter and have less inertia, giving less unsprung weight and less rotating inertia, something which is rarely achieved in chassis development.

The next step was to use unequal piston diameters. Kawasaki said that the pressure distribution was not equal when the disc was moving and having a smaller piston at the leading end of the pad evened it up. Others said that the trailing half of the pad ran hotter, making the coefficient of friction drop and needing higher pressure – again the smaller diameter went on the leading edge. Another school of thought was that as the trailing edge ran hotter it would wear more, and to prevent uneven wear it was necessary to put the smaller piston on the trailing edge. Nissin, AP, Kawasaki, Suzuki, Yamaha and Honda put the smaller diameter piston at the leading edge. ISR kept four individual pads, one for each piston, and used equal diameter pistons.

Figure 6.8 (a) and (b) Four-piston ISR caliper has equal diameter pistons and separate pads

The various factors used in equation 6.5 show how the brakes may be uprated but it will also be necessary to allow for the consequences of this. More brake power will mean more heat generated at the pads which could overheat the fluid, the pad materials or the disc material.

To counter this it may be necessary to provide better pad/disc cooling (see Chapter 7, Aerodynamics) or to alter the mass or the surface area of the disc (see below). The running temperatures can be tested by making a thermocouple to run into the brake fluid through the bleed port or by using temperature sensitive paint/stickers which are available through AP racing. (See Chapter 11 and Appendix.)

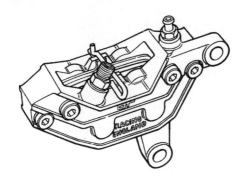

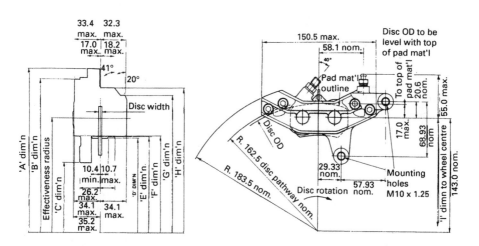

Figure 6.9 AP Racing supply full data sheets for their components, such as this unequal size four-piston caliper. Racing calipers are usually available in either aluminium alloy or magnesium alloy. (*Automotive Products*)

An increase in the brake force will also change the bike's attitude under braking and may change its tendency to lock the wheel or overturn. If the centre of gravity is known, this can be calculated, as shown in Figure 6.19.

Figure 6.10 Tokico and Nissin supply four-piston calipers as OEM for the Japanese manufacturers.

As a first approximation:

$$R_f w = mg(w - cgx) + F(cgy) \qquad (6.6)$$

$$R_f + R_r = mg \qquad (6.7)$$

Where R_f is the weight on the front axle, R_r is the weight on rear axle, w is the wheelbase, m is the mass of machine (mg = weight of machine), cgx is the distance of centre of gravity behind the front wheel spindle, cgy is the height of the centre of gravity above ground, and F is the braking force at the tyre (from 6.5).

If $F > \mu_w R_f$ (where μ_w is the coefficient of friction between the tyre and the road) then the tyre will slide and the limiting force $F_{max} = \mu_w R_f$.

If $F < F_{max}$ and $F > mg.cgx/cgy$ then the machine will overturn (at which point $R_f = mg$). The spreadsheet program in the appendix calculates F for given input forces and brake geometry, and calculates this limiting condition.

The weight transfer will compress the front suspension and extend the rear suspension which will, in turn, alter the height of the centre of gravity, cgy (and to a much smaller extent, its position, cgx). A second approximation must take this into account (the force/spring deflection may also have some influence on the choice of spring rate and anti-dive mechanism). The force fed into the front spring(s) will be $(R_f - R_{fo})\cos\theta$ where θ is the angle of the forks from the vertical and R_{fo} is the static load on the front axle.

Note that if brake torque is applied to the sprung end of the suspension then this force has to be subtracted from the total.

This will compress them a distance of $(R_f - R_{fo}) \cos \theta / S$, where S is the total spring rate (i.e., both springs) and will lower the sprung end of the fork by $(R_f - R_{fo}) \cos^2 \theta / S$. It will also shorten the wheelbase (w) and the position of the centre of gravity (cgx) by an amount $(R_f - R_{fo}) \cos \theta \sin \theta / S$.

Pure calculation gets too unwieldy at this point and it is far easier to use a scale drawing to find the deflection that this causes at the centre of gravity.

The second approximation needs to calculate the weight transfer and overturning torque using the new centre of gravity position. Although the calculations are straightforward, they are tedious to repeat for different values and this is one advantage of the spreadsheet as it will re-calculate everything each time one of the values is changed.

This will be accurate enough for most applications, but to get a more accurate, third approximation, aerodynamic forces have to be added. This is obviously more significant at high speeds. It affects the weight distribution and the weight transfer in the following way:

$$R_f + R_r = mg - lift \tag{6.8}$$

$$R_f w = mg(w - cgx) + F(cgy) - drag.cpy - lift.cpx \tag{6.9}$$

Where *drag* is the aerodynamic drag force, *lift* is the aerodynamic lift force, *cpx* is the distance of the aerodynamic centre of pressure behind the front wheel spindle, and *cpy* is the distance of the centre of pressure above ground.

It is only possible to get this data from tests or to estimate it, assuming, for example that $cpy \approx cgy$ and that lift tends to zero, until proven otherwise.

The weight of the caliper is significant. It is unsprung weight and has to move with the steering, so any reduction in its mass or in its distance from the steering axis will be an improvement. Calipers used to be fitted to the front edge of the forks – which is about as far as they could be from the steering axis and still be bolted to the fork leg. Nearly all designs now put them behind the fork leg, virtually on the steering axis.

A single disc and caliper is to be preferred to twin discs, unless the extra braking capacity is essential. The caliper is usually bolted directly to the fork leg or to a carrier plate, in either case the bolts are in single shear.

The brake force between caliper and disc is $F_2 \mu_B$ and, if the disc radius is approximately half of the wheel's rolling radius, then this force will be in the order of two times the machine's total weight. It will be shared approximately equally between two mounting bolts. If M10 bolts are used, the cross-section area of each will be 78.5 mm² (0.12 in²) so a machine weighing 500 lbf total will typically put a shear force of 500/0.12 or 4166 lbf/in² on the bolt. Using M8 bolts would increase the shear force to 6410 lbf/in², which is why the bolts need to be good quality, in good condition, a good fit in the fork lug and done up tightly.

This assumes that the caliper is mounted so that the force is distributed evenly between the two mounting bolts. If it is mounted on a plate then the plate has to be able to take the same shear loadings and it may position the caliper so that its anchoring force is no longer directed at the two mounting bolts. In this case the shear stress in one may be greatly reduced, and the shear stress in the other will be greatly increased. This will be the case if a tangent drawn from the centre of the pad area does not go in between the two mounting pins (see Figure 6.6).

6 Pads

The pads are the most easily changed parts of the brake system and the choice will be original equipment, with possibly a competition option, or one of several aftermarket manufacturers. High friction values are not particularly important because the same brake force can be generated with lower friction and higher line pressure or greater piston area in the calipers.

What is more important is that the pads retain their performance over the full temperature range and speed range which the bike is likely to achieve, and this can only be discovered by testing. The wear rate is important on a road bike and so is the pad's consistency as it wears down. As the pad becomes thinner it is less effective as an insulator and will conduct more heat from the disc to the caliper and the brake fluid.

The local pressure between pad and disc can be increased by reducing the pad's contact area; machining grooves in it will do this and will provide channels to clear debris. Both will help brakes which have to work in dirty conditions and may prevent brake squeal. A word of warning: pad material may contain asbestos – wear a mask, do not blow the brake clean with an air line, damp down the dust.

New pads are often chamfered, reducing the initial contact area, to speed up the bedding-in process, while the grooves sometimes found in pads are usually wear indicators or are there to stress-relieve the material to prevent cracking under thermal stress. Pads can be bedded in by moderately hard use, as long as they are allowed to cool down between operations (light, continuous use is more likely to cause overheating and 'glazing' – see below).

The materials used in pads tend to be:
(a) Organic.
(b) Organic with metal particles.
(c) Sintered metal.
(d) Ceramic or ceramic/sintered metal.
(e) Carbon fibre (used with carbon fibre discs).

The manufacturers usually supply data sheets which list the material's properties, particularly the variations in friction with temperature and rubbing speed which are essential for calculating the basic geometry such as disc diameter, piston diameters, line pressure, etc. The attractions of ceramic materials are that they have good anti-scuff properties, do not lose strength at high temperatures, and do not transmit heat too efficiently (i.e., from disc

Figure 6.11 Wet braking test rig built by LEDAR. (See Chapter 11)

surface to brake fluid). The pad material must be compatible with the disc material – the pad manufacturer will supply this information.

Tests which we made in 1978 showed that pad material had the biggest influence on brake performance in wet conditions (see Chapter 11). Basically, as more water is delivered to the pad, the force between it and the disc gradually decreases, until at a certain level the grip suddenly diminishes and the performance deteriorates drastically, at least it did with the OEM compounds used in the 1970s. If others show this characteristic, they do not do it in the normal range of pressures and water flow rates found in everyday use. The tests were run on different disc materials, with slotted and drilled discs, different pad configurations and different line pressures but it was only the pad material that made any significant difference to the performance.

The coefficient of friction for pad materials lies in the range 0.3 to 0.5. Competition pads tend to be at the lower end of this scale but they also tend to vary less over wide ranges of temperature and speed. A constant level of friction would be ideal for bike applications, but most materials have a coefficient which gradually falls as it gets into high temperature regions. This is because in cars there could be a slight mismatch left-to-right which would make the car pull to one side. However, the brake doing the extra pulling will run hotter and if this then reduces its friction, the car will tend to stabilize automatically.

The disc/drum and lining should be smooth and polished over their full areas after bedding in. Overheating the lining materials will initially damage the surface material; this will be weakened and will quickly wear off during subsequent normal use. Very light, continuous use will produce overheating and may not tear away damaged particles from the surface, which then becomes clogged with a hard, shiny layer (called glazing). Severe braking is usually enough to break up this layer and wear it away.

Organic pad materials consist of fibres held in a resin consisting of carbon, hydrogen and oxygen. When this is consistently overheated, it burns off the hydrogen and oxygen, leaving carbon which is brittle and is quickly worn away, leaving nothing to support the organic fibres. During this process, which is known as 'fade', the coefficient of friction drops markedly yet the lever travel and pressure do not change.

The same applies to the linings used in drum brakes, with one difference. The friction between the drum and the leading edge of the lining tends to pull the lining away from its cam or piston, into harder contact with the drum. This self-servo action is made use of in drum brakes which may have two or four leading shoes (i.e., the cam operates the leading end of the shoe). It depends on the geometry of the cam and shoe and it increases exponentially with the coefficient of friction of the lining. If there is too much self-servo action the brake will grab and will be difficult to control. It can be reduced by using lower friction linings or by chamfering the leading edge of the lining so that it comes into contact with the drum further away from the cam. See Figure 6.22.

7 Discs

The effective diameter controls the brake torque and, with the wheel size, determines the force generated between the tyre and the road (see equation 6.5). Assuming that there is uniform braking over the full working area of the disc, the effective diameter will be midway between the outer and inner edges of the working area. This is good enough for most calculations but may not be strictly true as there is a difference in the rubbing speed between the inner and outer edges.

The other considerations are its strength and weight . Because the disc turns with the wheel its inertia is important to the overall performance of the machine, so any reduction in weight is worthwhile but an increase in radius not only makes the disc heavier, it also increases its radius of gyration. The use of four- and six-piston calipers permits a high effective radius with a narrow (therefore light) working area.

Strength, particularly under thermal distortion, was a problem when discs were rigidly bolted to their carriers. Very thick, but hollow, ventilated discs provide one answer, giving good strength and cooling.

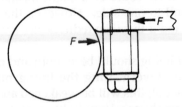

Figure 6.12 The caliper mounting is usually a bolt in single shear.

Another answer is to use a floating disc, as thin and light as possible, mounted on dowels so that it is free to move – and to expand and contract – a small amount axially and even to turn slightly relative to the wheel in some designs. Even so, some of the early types were prone to overheating and cracking.

Whether the disc is bolted directly to the carrier or whether a web is bolted to the carrier and holds the disc on dowels, the bolts are highly loaded, in single shear and fatigue. Use the correct bolts, at the correct torque and do not re-use them if they have become worn or damaged. Use a tab washer, thread lock or locking wire but do not use scratched or corroded bolts. The bolts should be high tensile (the overall shear force is greater than at the caliper mountings, although it is shared between more bolts) and the shouldered portion or the plain shank should be a good fit in the disc/carrier.

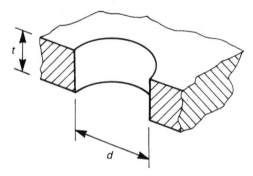

Figure 6.13 Drilling holes in discs not only makes them lighter but changes the surface area according to the proportions of the hole diameter d and the disc thickness t.

Drilling (and slotting) discs can be used for several effects: to make the disc lighter, to alter its surface area and to help it break up and clear away debris at the pad surface.

Of these, making the disc lighter probably has the most value, and any hole will remove material and weight. Obviously it is important that the holes do not weaken the disc – drilling holes close together on the same radius or in a radial line, would not be a good idea.

As the holes pass under the pad, the contact area changes continuously, so the local pressure varies and this is probably good for keeping the pad clear of any glazing or other contamination on the surface layer.

If the disc weighs less it will heat up faster and, for a given heat input, will reach a higher temperature. The holes can also make the surface area of the disc greater, smaller or leave it at the same value. As the disc cooling depends to a large extent on surface area, a careful choice of hole size can help it to run hotter or to cool it.

If the thickness of the disc is t and the diameter of the hole is d, then the circular area of the hole is $\pi d^2/4$, removed from each side of the disc. The wall area of the hole is πdt, added to the disc, so the total change in surface area is $\pi ndt - \pi nd^2/2$, for n holes.

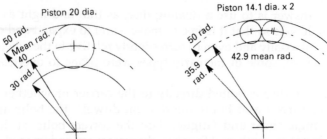

Piston 20 dia.   Piston 14.1 dia. x 2

50 rad.
Mean rad.
40
30 rad.

50 rad.
35.9 rad.
42.9 mean rad.

Figure 6.14 (a) The advantage of four-piston calipers is in being able to use lighter discs. If the disc keeps the same maximum radius of 50, and the calipers are chosen to give the same piston area then the four-piston type gives a greater mean disc radius, and greater braking torque by 7.2 per cent for the same force between pad and disc. It also allows a narrower disc, by 24.3 per cent, so the weight will also be reduced by this amount. Strictly, to get a mean radius so that there is equal pad area to either side of it, it is necessary to take root mean square values and the effective mean radii become 41.2 for the two-piston caliper and 43.5 for the four-piston layout. The increase in brake torque will then be 5.6 per cent.

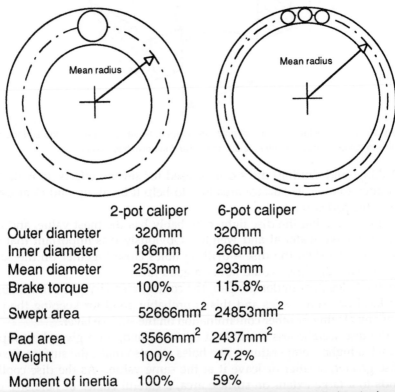

Mean radius

Mean radius

	2-pot caliper	6-pot caliper
Outer diameter	320mm	320mm
Inner diameter	186mm	266mm
Mean diameter	253mm	293mm
Brake torque	100%	115.8%
Swept area	52666mm^2	24853mm^2
Pad area	3566mm^2	2437mm^2
Weight	100%	47.2%
Moment of inertia	100%	59%

Figure 6.14(b) The four-piston arrangement in Fig 6.14(a) gives the same piston/pad area and wear as the two-piston type and is optimized for road use. This shows a six-piston caliper optimized for racing, where the inertia of the disc and braking power take precedence over wear rates. For a given maximum disc diameter, the weight and inertia of the disc can be roughly halved. Because the swept area, weight and the pad area are also reduced, the wear rate (pad and disc) will be increased and the brake will run at a higher temperature, getting up to temperature more quickly

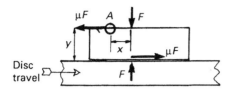 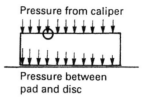

Pressure from caliper

Pressure between pad and disc

Disc travel

Figure 6.14 (c) If a pad, pushed against a disc, is anchored at a point A as shown, then the forces on the pad set up a torque $T (= \mu Fy - Fx)$ which tends to increase pressure at the leading edge of the pad and reduce it at the trailing edge. F is deemed to act through the centre of pressure of the pad. To get even pressure distribution it is necessary to make $Fx = \mu Fy$. Note that x, as shown here can be positive or negative and that the torque on the pad can be positive, zero or negative (increased pressure on trailing edge) depending on the dimensions of the pad and its retaining pins. Using long pads with unequal diameter pistons alters the initial pressure distribution, effectively shifting the centre of pressure towards the larger piston

If $d < 2t$ the surface area will increase.
If $d = 2t$ the surface area will stay the same.
If $d > 2t$ the surface area will decrease.

The change in weight will be proportional to $\pi n d^2 t/4$ multiplied by the density of the material.

The material favoured by the Japanese for road bikes is one of the stainless steel alloys, whereas practically every other manufacturer prefers cast iron. In racing applications where cost and wear are not high priorities, armoured aluminium alloy, titanium (!), and carbon fibre have all been used. A rough guide to the properties of some of these materials is shown in Table 6.3.

Given that cast iron is cheap, easy to produce, has good resistance to scuffing and, while it is not a good conductor of heat it can withstand thermal shock, it does have a fair amount to recommend it, apart, of course from rapid corrosion.

Typical disc runout figures are under 0.15 mm (0.006 in) on a 5×254 mm diameter disc (0.2×10 in). While zero runout is ideal, if the hydraulics make the brakes drag (as, for instance, on the Yamaha RD350 models) then a small amount of runout (0.10 to 0.15 mm; 0.004 to 0.006 in) is desirable. It will knock the pads away from the disc, reducing drag and avoiding the risk of overheating or glazing the pad surface.

If the total pad clearance is known, the necessary lever travel can be found from equation 6.4 above, or by using the spreadsheet program in the Appendix.

8 Anti-dive

This is an interaction between the brakes and the front suspension which *either* prevents the springs from compressing *or* makes them compress at a slower rate. The distinction is between the *distance* compressed and the *speed* at which the suspension compresses when the brake is applied.

Table 6.3 Disc materials

	Cast iron	Steel	Stain- less	Aluminium alloy	Titanium alloy
Density, g/cm^3	7.2	7.9–8.7	7.9–8.7	2.6–2.9	4.5
Hardness, HV	200–250	700–1000	130	armoured	armoured
Thermal conductivity, W/m–K	58	26–58	14	70–240	15.5
Specific heat, kJ/kg–K	0.5	0.42–0.49	0.5	0.94	0.47
Tensile strength, N/mm^2	240–400	400–1000	400–1000	200–300	300–900
Young's modulus ×1000 N/mm^2	121–175	206	206	70	110
Melting point, °C	1200	1450	1450	480–655	1670

Calculations such as equation 6.6 and the spreadsheet program (see Appendix) show that there is an optimum height for the centre of gravity, at which maximum braking force can be found. The amount of dive can sometimes be regulated so that the centre of gravity is at the optimum height during braking.

When the brakes are used and full pressure is reached gradually, then the weight transfer is also gradual. If full pressure is applied immediately then the sudden and rapid compression of the front suspension can make control difficult. A temporary increase in spring stiffness or in compression damping can be used to take the violence out of this pitching motion, simply to make the rider more comfortable and give him more control, regardless of the change in ride height.

There are four main types of anti-drive:

(a) Feed brake torque into suspension to regulate the ride height and so control the centre of gravity height.
(b) Load sensitive suspension, which has some control over the ride height during braking and also controls the rate at which the suspension compresses.
(c) To use the brake in some way to alter the compression damping, purely to reduce the rate of dive.
(d) To route inertia forces directly to the unsprung part, avoiding the suspension (see Chapter 5).

114

Type (a) comes in many forms. Some leading- or trailing-link forks have it built in as an inherent part of the design. The phenomenon appeared on a Honda moped built during the early 1970s, which had a pivotted-link front suspension that actually *rose* during braking, so obviously the force in the

Figure 6.15 Brake evolution.
(a) This eight-leading shoe drum brake (there is a similar backplate on the other side) was fitted to the Dresda Triumph which won the 1970 Barcelona 24-hour race. **(b)** Early, rigidly mounted discs suffered thermal problems; this ventilated disc was used on a Honda CB1100R.

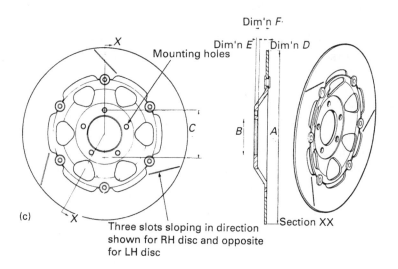

(c) A better solution was to use lighter rotors, located by dowels on a central carrier, giving the disc sufficient freedom to float and to expand. (*Automotive Products*).

(d) Grand Prix bikes (this is a 1987 Honda) took it a stage further and used carbon fibre discs for lightness and the material's ability to cope with the thermal loads.

brake anchor was greater than the weight transfer. The idea was incorporated into some leading link forks which we built for a 125 racer but which were not very well executed and tended to lock the suspension completely when the brake was used. Others tried different approaches with rather more success. There was a BMW endurance racer, followed by Kawasaki's GP racer, the Ron Williams Honda Britain F1 racers and Peckett & McNab machines, all of which had similar torque linkages. The torque is transmitted via an adjustable lever so that the required force is fed into the sprung end of the suspension.

The frictional force between pad and disc is $F_2\mu_B$ (see equation 6.3), so the restraining torque on the caliper must be $F_2\mu_B r_B$, where r_B is the effective radius of the disc. In the type of linkage shown in Figure 6.16, the force in the first strut will be $F_2\mu_B r_B / r_1 \sin\alpha$, where r_1 is the length of the torque arm from the wheel centre and α is the angle between this and the strut. A similar calculation has to be made for successive levers in order to arrive at the force which is directed into the top of the suspension. Again, this is something which can be recalculated swiftly for different values if a spreadsheet is used.

Type (b), load-sensitive suspension is employed on many road and race bikes. As the compression of the spring increases, the spring rate also has to

116

increase (i.e., a rising rate – see Chapter 5). Progressively wound springs and air-assisted forks do this naturally. The cumulative spring rate has to be known in order to work out the deflection for a given load. Increasing the level of oil in the forks increases their compression ratio and gives a steeper

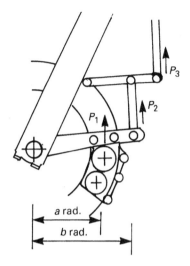

Figure 6.16 Mechanical anti-dive. The restraining force on the caliper can be stepped down (or up) by a succession of levers. In this case $P_2 = P_1 a/b$ and so on, until the required force P_n is fed into the sprung end of the suspension.

Figure 6.17 A road going version of P & M's F1 machine. The pivoted leaver has a series of holes so that the anti-dive force can be adjusted. This is on the highest setting

117

rising rate. Some forks have a load-sensitive damper. When the deflection reaches a certain point, a valve is closed, giving a large increase in compression damping.

Type (c) has several variations:

(i)   To use hydraulic line pressure to open/close a valve in the damping circuit, to give stiffer compression damping (Suzuki, Kawasaki and Yamaha).

(ii) To use a plunger to restrain the caliper so that the force of the caliper against it during braking would move the plunger and increase the compression damping (Honda).

(iii) To use the stop light switch to actuate an electrical solenoid which would move the necessary valves in the compression damping circuit (Kawasaki).

(iv) A combination of type (b) and (c) in an Australian patent application which had the caliper restrained by an hydraulic piston so that the brake force pressurized hydraulic fluid. This pressure was used to move two more pistons on top of the suspension springs, thus increasing their pre-load in proportion to the braking force.

Type (d) is inherent in some suspension designs, where a near-horizontal wishbone or link carries the weight of the bike and transmits brake/engine thrust. Inertia under braking appears as a horizontal force at the CG which is transmitted through the wishbone without passing through the spring or creating a vertical load.

Anti-dive in its various forms can achieve several objectives; it can control the height of the centre of gravity; it can prevent all the suspension movement being used up during braking; it can reduce pitching. The value of all this depends on the bike and on the surface conditions. If stiffer springs or damping give better control – as they might on a bumpy circuit – then this type of control is worthwhile. If the bike has a problem of locking its front wheel, then raising the centre of gravity during braking will allow more brake force to be used (and so would a forward shift of the centre of gravity or the use of a tyre with more grip). A mechanical type of anti-dive can raise the centre of gravity under braking (or prevent it from being lowered too much). If the bike is able to lift its back wheel under braking then the only improvement will come from lowering the centre of gravity (or moving its rearward), in which case *less* anti-dive would be necessary.

The value of anti-dive as a means of increasing braking power was demonstrated during some speed tests on MIRA's timing straight, which had a stopping area of less than 300 yards. The fastest machine which we had run there was a Suzuki GSX-R1100, which had stopped with some room to spare from a speed of 153 mph and for most of the braking distance the rear wheel had been about an inch clear of the ground. The P & M Kawasaki 1100 shown in Figure 6.17 stopped equally comfortably in the wet from 143 mph. If

Figure 6.18 This anti-dive system was built by Barry Schultz in Australia, around 1980. The caliper torque is fed into an hydraulic cylinder, with a slave cylinder mounted in the top of the fork leg so that the piston can push down on the fork spring, raising the pre-load during braking.

the Suzuki stopped at 1 g (which it must have been close to), the P & M stopped at 0.87 g – on a wet track.

Obviously, on a wet track it was more likely to slide the front tyre than to lift the rear, so anything which kept the centre of gravity high up would help the braking performance.

The graph in Figure 6.19 shows how critical it can be. This is taken from figures for a Honda CR500R which was being modified to run in hill-climbs, where acceleration and braking are very important. The centre of gravity was measured (see Chapter 2), and with some assumptions about tyre grip, the figures were fed into the spreadsheet shown in the Appendix, varying the height of the centre of gravity and then increasing the hand pressure until the program indicated that the wheel would lock or the bike would overturn.

From these limiting figures, the computer obligingly drew the graph in Figure 6.19 which shows how critical the height is. This is the centre of gravity height during braking and, as the program also shows the axle load, it is possible to work out the spring rate and static ride height which will give the desired result – or at least get close to it. The outcome would have to be confirmed by track tests and it looks like two alternative set-ups would be needed: one for grippy tracks and one for wet or poorly-surfaced tracks where a high centre of gravity would be more beneficial.

119

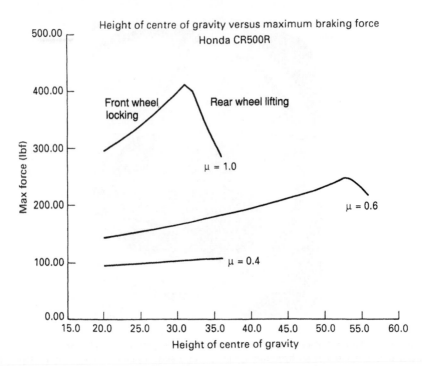

Figure 6.19 The height of the centre of gravity controls the maximum front wheel braking force. This shows the effect of changing the centre of gravity height on a CR500R Honda modified for hill-climb racing. It is obviously important to arrange the suspension to keep the centre of gravity close to 32 in during braking in the dry. In the wet, the higher the better. The spreadsheet program in the Appendix will produce results like this very quickly. Moving the centre of gravity forward or shortening the wheelbase will give different values for the optimum height.

This exercise shows a couple of interesting things. First, how critical the centre of gravity position is. An error of 2 inches one way would lose about 80 lbf from the available braking force – a 20 per cent reduction. Getting it 2 inches wrong in the other direction would be less serious, but it would still cost about 30 lbf or 8 per cent of the total *and* an error in this direction would make the bike more prone to lock up the front wheel. In this case the rear brake could be used to augment the front but with only 8% or so of the machine's weight on the back wheel, it would take a lot of skill to control. When friction is reduced – in the wet – it becomes essential to use the rear brake as well as the front; this can add about 50% to the total braking effort.

Second, would a test rider pick up information as clearly as this? Some machines feel noticeably better than others under heavy braking but I have never felt, nor have I heard any other rider comment, that a bike would be better if it were an inch or so higher/lower. Now that the critical point has been located, it will be relatively easy to start high, run tests and gradually lower the bike until the rider (and hopefully the results) reach an optimum level.

120

Third, because the Honda is a very tall machine, with some 12 inches of suspension travel, the first instinctive reaction is to lower it as far as possible and start the tests from there. Clearly this is not necessary. The centre of gravity height suggested by the program is not necessarily the ultimate. These results would be altered by moving the centre of gravity backwards or forwards as well, and the final geometry will be decided by the centre of gravity position for acceleration (using the program RL.BAS) and then choosing front suspension to get the optimum height under braking – with a measure of anti-dive if necessary.

In fact, it will be necessary to recalculate the whole thing anyway because this is all based on level ground performance. If the bike is on a slope at an angle θ to the horizontal, going uphill, the forces R_f and R_r are now the axle loads at right angles to the road surface:

$$R_f + R_r = mg \cos \theta$$

If the brake force $F > \mu_w R_f$ then the tyre will slide, as before

and $F_{max} = \mu_w R_f$.

The bike will overturn if:

$$F < F_{max}$$

and $F > mg (cgx \cos \theta / cgy + \sin \theta)$

For the downhill case, substitute $360 - \theta$ for θ (which makes $\sin \theta$ negative).

Whether a bike needs anti-dive or not depends entirely on these equations. Grand Prix bikes stopped using it because they can already overturn and they need to lower their centre of gravity in order to improve braking performance, at least in the dry. Many other bikes simply follow the fashion set by GP racers when, in fact, they could benefit from anti-dive. Any bike, from dirt bike through to road bike, which is capable of locking its front wheel in certain conditions will benefit from anti-dive in those conditions.

9 Pro-squat

Pro-squat is the complement of anti-dive. It is brake torque fed into the rear suspension so that the rear springs compress. It is used on some cars (along with anti-squat, which raises the rear suspension during acceleration) but is of minimal use on bikes because with their short wheelbase and high centre of gravity, they are capable of lifting the rear wheel during braking. There-fore, the rear wheel cannot contribute to the braking force and cannot affect suspension movement. The brake is arranged so that the force in the torque arm tends to compress the spring or tends to pull the sprung part of the bike downwards. This is of limited use because the rear brake is only effective if the rear wheel is carrying some load and in this case the limiting condition is that the front wheel will lock. Lowering the centre of gravity, which is what pro-squat will do, will only reduce the brake force needed to make the front wheel lock. Anti-squat brakes would raise the centre of gravity when the rear brake was used and would therefore allow more front brake – if the rider were good enough to be able to modulate front and rear in this way. It is worth mentioning if only because some 1987 and 1988 works Honda GP bikes had a complicated bell crank linkage which could only raise the rear suspension when the rear brake was used.

Figure 6.20 (a) and (b) Rear brake force can be anchored to the sprung part of the bike in such a way that it compresses the rear suspension (pro-squat) or, in this case, extends it. Honda use this linkage on the VRFR750R (a) and on some of the GP racers, while others, such as this 1987 model, with carbon fibre disc, had a linkage arranged to produce anti-squat (b)

10 Anti-lock brakes

It is worth considering the anti-lock braking system (ABS) because it already exists in several forms and its potential in cars is quite formidable. Whether it will be better than an expert driver is no longer the point because the anti-lock braking system already has the advantage that it can select one brake (or more) where the driver has only one pedal. In bikes it has the disadvantage that it cannot distinguish between braking force and cornering force and the first crude versions available (early 1989) were only of use when the bike was upright plus or minus a couple of degrees.

Later versions are more refined, weigh less and the dump-reapply cycle is more rapid, with a less pronounced on-off effect. Systems developed by BMW with TAG Kugelfischer and by Yamaha have certain similarities. Speed sensors on each wheel send data to a central processing unit which, when it decides that a wheel is decelerating too quickly or is decelerating at a significantly different rate to the other one, brings the ABS into play.

NORMAL BRAKING OPERATION

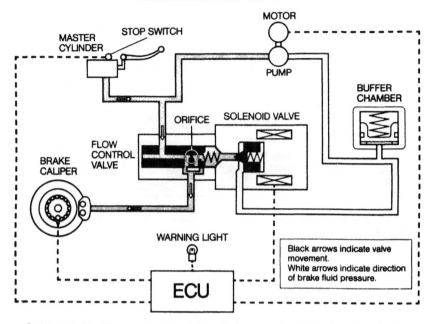

During normal braking operation, the master cylinder pressurizes the brake caliper directly.

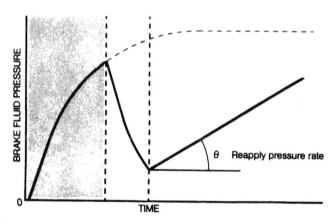

*Front and rear system operation is identical although only one system is shown.

Figure 6.21 (a), (b) and (c) Schematic of Yamaha's ABS showing the three phases of operation – (a) brake applied, (b) line pressure released when the processor decides a wheel is locking and (c) pressure is reapplied at a predetermined rate. *(Yamaha)*

On the Yamaha system a valve is opened, dumping brake pressure to a buffer chamber in which there is a piston with a spring behind it. This releases the brake at a predetermined rate and a pump recycles fluid around the circuit, maintaining lever or pedal pressure with very little movement to let the rider know the system is operating. The processor monitors the wheel speed 125

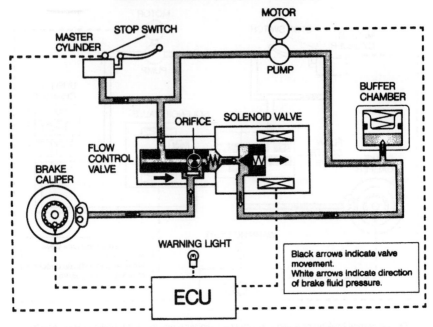

As the ABS is activated, the solenoid valve is opened releasing brake fluid pressure to the buffer chamber. The flow control valve moves due to the pressure difference before and after the orifice. Movement of the flow control valve cuts the passage to the caliper before the orifice. Farther movement opens the passage after the orifice allowing the caliper to depressurize.

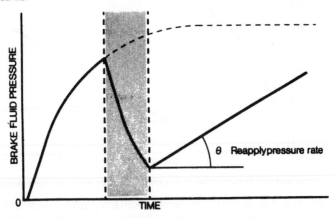

Figure 6.21 (b)

times a second and, having dumped the line pressure, the flow control valves can reapply it at a carefully chosen rate.

If the wheel starts to lock again, then the dump-reapply cycle continues, at up to ten cycles per second in very bad conditions (i.e. very low tyre traction or excessively high line pressure). In borderline conditions, one cycle is often

124

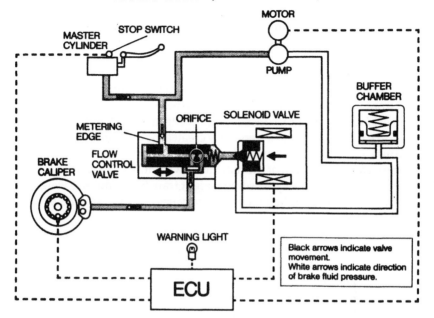

The solenoid valve closes and the caliper is repressurized. Brake fluid pressure is controlled to a specified amount by the metering edge.

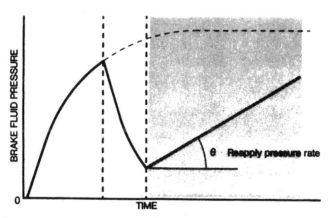

Figure 6.21 (c)

enough to prevent the wheel locking: the rider feels the lever or pedal click and is able to regulate the pressure himself to avoid further ABS operation. In these conditions the system will work even when the bike is cornering at moderate angles of lean, but the system itself cannot detect cornering forces and will not prevent a wheel sliding if the tyre is pushed too far.

ABS has two uses: first, in very slippery or sudden emergency conditions when the rider would not normally have the skill or the time in which to find optimum braking levels. Second, in normal use (especially where surface conditions are changing) it allows the rider to quickly find the maximum braking level and to hold the machine just below this limit, without fear of locking the wheel if he should go too far or if the surface conditions should deteriorate.

The system is of more use on machines which have a long wheelbase and low CG because, even on good road surfaces, their limiting condition will be that the front wheel will lock. This is because there is less weight transfer to the front wheel (see Figure 2.6); consequently the front wheel will lock at a lower level of force than a bike whose geometry creates more weight transfer. And, to achieve maximum braking, the rear brake must be used because the rear wheel still carries a fair proportion of the machine's weight. This requires a lot of skill because there will inevitably be some weight transfer, removing weight from the rear axle and making the rear wheel very easy to lock, so ABS becomes a useful aid at both wheels.

Sports bikes with a higher CG and shorter wheelbase transfer more weight to the front wheel during braking and in good conditions will transfer all the weight on to the front. This leaves the rear wheel incapable of making any braking effort and the limiting condition is that the bike will overturn, lifting the rear wheel off the ground. In this type of machine ABS would only be of use in slippery conditions when it would be possible to lock the front wheel and when rear wheel braking is necessary to achieve maximum stopping power.

When ABS is operating, the brakes are, in effect, let off and put back on again. No matter how rapidly this is performed, the brakes are still not in use for a certain proportion of the time. Therefore the stopping distance will be longer than if the brake were on continuously, but just below the pressure at which the wheel will lock. The advantage of ABS is in the speed with which the maximum brake force can be found because the rider knows he can apply high pressure quickly, using the pulses at the lever/pedal to warn him when he has exceeded the limit, and allowing him to regulate the pressure and balance the front-to-rear braking, to keep the brake force very close to the limit, even in very slippery or highly variable conditions.

Drum brakes

From a performance point of view, the disadvantage of drum brakes is that they are heavy and, when they are built on a large diameter, they force large increases in the inertia of the wheel hub. In cast wheels, thermal expansion causes distortion because the hub is much more rigid at the root of each spoke than the free drum. Their advantage is that they can be operated by a simple cable or rod, small input forces being possible because of the drum's self-servo action.

Figure 6.22 shows the principle, using a small element of a brake shoe pressed against the rotating drum by a force F acting about the pivot point A.

If the drum rotates clockwise, the frictional force between it and the shoe will be clockwise on the shoe, anti-clockwise on the drum. The moment of

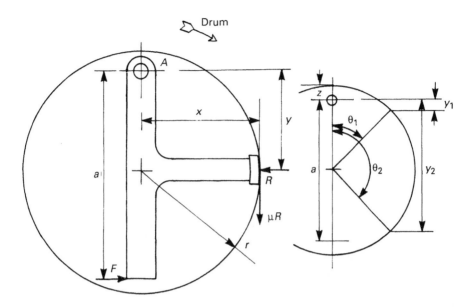

Figure 6.22 Principle of a drum brake, showing essential dimensions and forces acting on a small element of the shoe.

all the forces taken about the pivot point A will be zero:

$$Fa = Ry + \mu Rx$$

Where R is the reaction of the drum against the shoe and μ is the coefficient of friction between the two.

So, $R = Fa/(y + \mu x)$

This is the case when the operating force acts at the trailing end of the brake shoe (called trailing shoe operation).

Now if the direction of the drum is reversed and made anti-clockwise, the frictional force is also reversed in direction. The equation becomes:

$$R_1 = Fa/(y - \mu x)$$

For all positive values of x, R_1 is greater than R, i.e., the reaction at the drum and the brake force is greater when the activating force is applied at the leading end of the shoe (leading shoe operation). Consequently drum brakes fitted to bikes usually have cam operation at the leading end of each shoe and some designs had four or even eight shoes.

If the radius of the drum is r, the pivot point is distance z from the drum and the shoe extends for an arc θ_1 to θ_2 referred to the pivot, then the full braking force will be the sum of all the elements like that shown in Figure

127

6.22, over this arc. The position of each element relative to the pivot (x, y) will be:

$x = r \sin \theta$
$y = r(1 - \cos \theta) - z$

The total braking force for each shoe will be:

$$\Sigma \, \mu R \; = \int_{\theta 2}^{\theta 1} \frac{\mu F a \; \cdot d\theta}{r(1 - \cos \theta) - z \pm \mu r \, \sin \theta}$$

Note: the plus or minus depends on whether the shoe is trailing or leading.

Manufacturers tend to rate brakes by a 'brake factor' which is, essentially, the brake force divided by the input force and takes the self-servo effect and coefficient of friction into account. For drum brakes the brake factor varies between 1 and 5, increasing exponentially with the coefficient of friction. Too much self-servo action makes the brake difficult to control, especially at low speeds, and tends to promote brake fade. It can be reduced by using linings with a lower coefficient of friction or by reducing the arc of contact of the shoe.

Chapter 7

Aerodynamics

The aerodynamic force on a vehicle increases as the square of its speed. Because power is the product of force and speed, the power requirement goes up as the cube of the speed. In other words, if it takes 4 bhp to reach 50 mph then to double the speed to 100 mph it will take $(2^3 \times 4) = 32$ bhp and to treble the speed to 150 mph will require $(3^3 \times 4) = 108$ bhp.

It is not quite as clear cut as this because some of the power is needed to overcome rolling resistance, drag in the driveline and brakes, and inertia of the moving parts of the drive. Some of these features are approximately constant, some vary in proportion to the speed. But as the speed gets higher, aerodynamic forces become dominant. There is a point where it requires so much more engine power for such a small increase in speed that it has to be easier to reduce the drag rather than look for more power. This is probably true at all speeds, except that it is more pronounced when you are looking for 100 mph from a 25 bhp engine than when a 50 bhp motor is available. In addition, it is not easy to guarantee reductions in drag when there are restrictions on the size and shape of the machine and limitations – both practical and legal – on the amount of streamlining that may be used.

When a body pushes through the air it builds up a high pressure region in front, somewhere along the sides the flow breaks up into a turbulent wake and this leaves a low pressure region behind. The drag force is the difference between the high and low pressures, acting on the area projected between them – the 'frontal' area of the bike. There is also friction between the air flow and the body but this is very small at the speeds we are looking at and pressure drag is by far the largest contributor to the total.

The drag force will be higher if the pressure difference is greater, if the effective area is greater, if the speed is higher and if the density of the air is greater:

Drag force = $0.5 \, d C_d A v^2$

Where d is the air density, v is the air speed, A is the frontal area and C_d is a constant called the coefficient of drag which is a comparison of the slipperiness or otherwise of different body shapes (of the same area, travelling at the same speed in the same air density). At US standard atmosphere (p = 1 atm or 14.7 lbf/in^2, temp = 288 K) air density is 1.225 kg/m^3 or 0.0766 lbf/ft^3.

The air touching the vehicle is considered to be attached to it, and travelling with it, i.e., its velocity relative to the vehicle is zero. Moving further away the air velocity increases progressively (the rate of change is called the velocity gradient) until it reaches 'free' air speed. This region of increasing velocity is called the boundary layer.

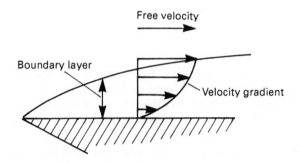

Figure 7.1 Air flow next to a surface sticks to it and travels at the same speed. Further away the air speed gradually changes (velocity gradient) until it reaches free air speed. This region of changing velocity is called the boundary layer

Separation is where the boundary layer becomes detached from the surface. The flow may be reversed immediately downstream of this point, leading to a large change in the pressure distribution which shows up as pressure drag. With high pressure at the front and low pressure at the rear, if the separation point is near the widest section of the vehicle, the turbulent wake will be wide and the pressure drag will be high. If separation can be postponed and the wake made narrower, then the pressure drop will act on a smaller area and the drag will be less.

The drag force acts in the opposite direction to the vehicle's movement. It is also possible to set up a pressure difference between the top side and the underside of the vehicle. This causes lift – either upwards or downwards (negative lift). A pressure drop between one side and the other will encourage the vehicle to move in the direction of the pressure. However, this force acts on the side area and if the centre of pressure is behind the centre of gravity then the vehicle starts to behave like a weather vane and turns towards the high pressure side. A well-balanced vehicle will be self-correcting because the pressure force will make it move, say, to the left while the couple set up between the centre of pressure and the steering axis will make it steer to the left and roll to the right and the two will largely cancel one another out. If the centre of pressure is far forward or if there is a large side area – as it might be with the 1950s style dustbin fairings, then high pressure on the side will cause unpredictable combinations of roll and yaw, a possibly unstable situation.

If a vehicle is travelling at an angle to the airstream ('yaw') then the frontal area will include a side projection and, for a long, thin vehicle this will amount to a greater area than a square-on front projection, so for a given pressure-drop, the drag force will be greater. Force is equal to pressure multiplied by area, so in general the drag force can be reduced by:

130

1 Making the pressure drop smaller.
2 Making the area smaller.
3 Making the *apparent* area smaller, i.e. making the shape more slippery or reducing its drag coefficient.

1 and 3 go inextricably together. A thin aerofoil shape penetrates the air and causes the minimum turbulence or wake behind it (most fish and birds have evolved the same shape). A flat plate, of the same frontal area, behaves as if it is much bigger and creates a wide, turbulent wake – a bigger pressure drop acting on an apparently larger area. A stubby shape, like a bullet, also creates a bigger pressure drop, but only across its true area.

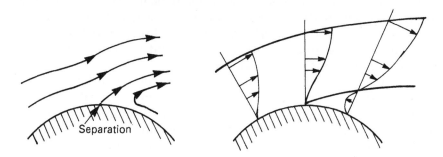

Figure 7.2 The boundary layer may separate from the surface, leaving a turbulent wake where the local air velocity may even be reversed

Parting the air, at ground vehicle speeds, is not difficult. Pointed fronts do not actually make a lot of difference until you reach several hundred miles an hour. Preventing separation and creating the smallest possible wake do make a difference, even at low speeds. The minnow shape is slipperiest, travelling blunt end first – the way the minnow tends to travel. The bullet, at motorcycle speeds, would be aerodynamically better if it travelled backwards.

This much was recognized long ago. Jenatzy built a record breaking car which reached 62 mph in 1899 and streamline principles were used by researchers and designers in the 1920s, like Rumpler, Dornier, Bugatti, Jaray, Klemperer, and in the 1930s Lange, Mauboussin, Lay, Kamm, Everling, Fishleigh, Heald and Reid. This led to designs like Bugatti, Tatra 87, Chrysler Airflow, Volkswagen Beetle, Lange and, later, the Citroen ID19, Porsche 356 and 911.

In all of these, the designer proposed to keep the air flow attached to the body for as long as possible, by using a gentle, rearward taper. The same thinking is still evident in recent concept cars and race cars like the Peugeot Oxia, the Porsche Carrera and 956, Peugeot 66, Mercedes C111, and General Motors 2002. But it did not meet the practical requirements of the three-box vehicle and once the necessary details like mirrors, radiators, uncovered wheels, door handles, etc. were added, it proved just as easy to take a more practical shape and optimize it aerodynamically. Thus while the concept car could have a C_d below 0.2, a production version would be around 0.3. A traditional three-box car (0.4 or more) could be sleeked down to 0.3 and still retain its practicality. The Audi 100 generation of cars was born.

So much about cars is necessary because so little work was done on bikes. Typically, faired bikes are about 60 per cent as efficient as cars. A car could have an area 1.66 times that of a bike and only create the same drag force. Record breakers took the familiar torpedo shape. Racers developed large frontal fairings in the 1950s and were promptly banned because the forward centre of pressure was thought to make them unstable with any amount of yaw. The dolphin fairing – little more than protection for the rider's torso and legs – became the norm and remained largely undeveloped (indeed largely unused) until production bikes reached 140 mph and racers were able to touch 160 mph. All along, various designs had recognized the principle that side and tail fairings could contribute most to efficient streamlining, yet it was never developed.

Since the early1980s, bike manufacturers have seen the value of an optimized shape – the pragmatic path chosen by Audi rather than the theoretically pure, teardrop shape – and have used it to improve performance, handling, rider comfort and to add another dimension to styling.

Originally, fairings were simply cowlings put around the front of an existing machine. They made it bigger and any gain in slipperiness was more than offset by the increase in frontal area. Then, as frames became the perimeter type, which went around the sides of the engine, it became easier to mount the fairing and to make the fairing an integral part of the machine.

It could carry headlamps, indicators, mirrors and instruments, taking a lot of mass and inertia off the steering – which improved handling considerably. It took wind drag – and weather – off the rider, improving comfort and reducing fatigue. As it no longer had to be much larger than the bike, it could reduce drag and it could lower the centre of pressure. This helped performance and it helped to remove lift at the front end of the bike – improving stability and handling.

From this point the whole bike could be optimized. There would be small ducts for radiators, deflectors to discourage too much air from going under the tank where it could only cause drag, and air scoops to feed cool, high pressure air to the carburettors. By 1988 a Ducati 851 Kit, which made only 86 bhp in still air, was able to reach 160 mph (the indications were that it was producing about 100 bhp when it was travelling at this speed). Compare this to the 1979 Honda CB 900 which made 83 bhp and reached 127 mph, or the 1980 Suzuki GSX1100E which, with a large aftermarket fairing, required 120 bhp to reach 150 mph.

As there is little chance of developing a fully streamlined machine (FIM regulations forbid streamlining beyond the positions of the wheel spindles and say that the rider's arms and legs must be visible from the side) then the only route is to optimize the existing shape. To put things into perspective, the sort of improvement which can be made is a gain of 1–3 mph at 100 mph when the mirrors are removed. This becomes a negligible increase at 50 mph. Conversely, it becomes significant and is worth several horsepower when the speed is raised to 150 mph. There are several well-known aerodynamic principles which can be applied to the detail design of any vehicle. A blunt shape at the front is not important. The first areas of concern are where the bodywork turns a corner or has a projection or a sudden change of shape. The front forks, the wheel itself, the edges of the screen and fairing, air scoops, the handlebars, the rider, all of these changes in shape can make the air flow break away from the surface and create the beginning of a turbulent wake. If this can be prevented (or delayed) or if the body further downstream can collect and smooth out the air flow, then the shape will produce less air resistance.

The front tyre is the first place to look. The top edge will be travelling twice as fast as the vehicle, and experiments suggest that the flow detaches at about 60° from the topmost point, i.e., at about the 10 o'clock position when viewed from the left side of the bike. A mudguard extended to just beyond this position will prevent this breakaway and will screen the high-speed top edge of the tyre. Ironically, this would be illegal under FIM rules, but if the mudguard is there to prevent spray and soiling, and if it did not extend more than 45° from the vertical, it would be allowed.

The same mudguard can shroud the fork legs and deflect air away from the engine region, to the side panels of the fairing. Leading edge radii need to be as large as possible. There is a critical point in the region of 1.5 inch – a radius of less than this is likely to cause separation. Some air flow is necessary to cool the brakes and this is almost certain to cause turbulence, but a low-pressure, swirling air flow behind the front wheel is not a bad thing – it prevents the full high-speed blast of air hitting the engine compartment – which would otherwise act like a large scoop. Also, low-speed turbulent air is more efficient at cooling the engine because more air is brought into longer contact with the hot metal. This is important for an air-cooled engine, but also works on liquid-cooled machines.

Where there are ducts for cooling air, to radiators or oil coolers, these should not add to the frontal area of the bike, and should not cause steps or projections which will make the air flow break away. It is better to take the air flow from the region behind the forks, where it will be travelling slower than the free air stream (consider the bike to be still and the air to be moving past it) duct it through the radiator and lead it to an exit in a region of lower pressure. This will do the least to cause break away at the front and will help fill in a low pressure area at the side or the rear, and so minimize the drag force. It will also avoid blowing hot air from the radiator on to other parts such as ignition components or carburettor intakes, which need to be kept as cool as possible.

Air intakes to feed carburettors, etc. need an uninterrupted flow of high pressure air, see Appendix, Table A.2. Again, they should not add to the frontal area or spoil its shape, and they should be of the smallest size necessary to provide the required air flow. If they are too big they will cause extra drag for no benefit in power.

The flow losses in pipes depend on the dimensions of the pipe. They are proportional to the length and are inversely proportional to the area. Where there is a bend, the loss coefficient goes above 10 per cent if:

$r/d \leq 1$ and $\theta > 40°$
$r/d \leq 2$ and $\theta > 45°$
$r/d \leq 4$ and $\theta > 60°$
$r/d \leq 6$ and $\theta > 90°$

Where r is the radius of the bend, d is the diameter of the pipe, and θ is the angle through which the pipe is bent.

So for a 90° bend, the radius of the bend should not be tighter than 6 × the diameter of the pipe in order to keep the losses acceptably small. A corrugated pipe tends to behave like a plain pipe of considerably smaller diameter.

The entry to the pipe should be well rounded – the loss coefficient can be as much as 50 per cent for a sharply cut-off entry (i.e., it will behave like a pipe of half the area).

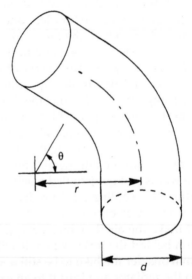

Figure 7.3 Flow losses inside pipes depend on the dimensions – the diameter d of the pipe, the radius r of the bend and the angle θ of the bend

Section changes in the pipe should also be gradual. In general, a diffuser (expanding cone) should not have an included angle of more than 15°. The loss coefficient will be proportional to the length and to the change in area. A nozzle (contracting cone) has a throttling effect, which causes the pressure to rise, but does not usually cause turbulence. Sudden steps, in either direction,

134

will cause turbulence and will generally reduce the flow capacity of the pipe. The pipe and its intakes should be matched to the air requirements of the engine (see Appendix), although the flow in the pipe will obviously vary with the bike's speed. For high speed operation, smaller pipes will deliver the required air flow for the minimum of drag.

The rider's body will certainly cause breakaway. It is possible that side and tail panels will persuade some of the air flow to re-attach, or to reduce the size of the low pressure wake. Tests made on trucks travelling in convoy showed that, as expected, the drag coefficient on the second and third trucks is reduced – by 30 to 60 per cent depending on the gaps between them. The drag coefficient of the leading truck could also be reduced by up to 15 per cent, simply by the presence of the following vehicle(s). So something which fills in the wake area – even if it is unconnected to the vehicle – can reduce its drag. Interference between vehicles can be felt at motorway speeds when, for

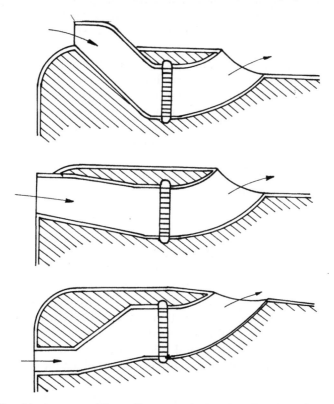

Figure 7.4 Possible ducts to a radiator. *Top*: worst, the intake scoop adding to the overall area as well as forming a sharp projection to cause turbulence. *Middle*: worse, a large, sharp-edged entry which will collect more air than is necessary and build up high pressure internally.
Bottom: better. A minimum-sized intake to provide just enough internal flow, located in a high pressure region but not so that it adds to the frontal area or breaks up the flow – the entry should be radiused. In the duct, the section is stepped out to reduce the velocity and avoid a pressure build-up at the radiator, and possibly cause some turbulence to get the maximum heat exchange from the radiator. The exit is smoothly blended into a low-pressure region

example, a large truck comes up close behind your vehicle (your vehicle speeds up) or when a large truck overtakes your vehicle (half way past, the interference will slow your vehicle; fully past, in the slipstream of the truck, your vehicle will accelerate).

Empirical analysis

The design of bodywork is usually governed by other requirements such as race regulations or the practical needs for things like mirrors. It is then necessary to make sure that the shape is no worse than it has to be. One problem is in testing, because the changes are often very small. Even with a wind tunnel there are difficulties in simulating the effects of moving ground and wheels.

To detect changes by road testing, it is essential that the weather varies less than the aerodynamic changes and this rarely happens. This difficulty can be eased by using a lot of test results, as these span the whole range of weather conditions and the spurious results tend to stand out from the mainstream.

For example, if you plot peak power against maximum road speed (or maximum air speed) for enough machines, it is possible to average out a graph which will then show whether any given machine is better or worse than the norm. There is a computer program in the Appendix which will do this. It is possible to see how machines have changed historically, or to compare machines of similar size, or to plot a graph for one machine with various levels of power output.

It is possible to produce a characteristic graph which matches measured performance and takes the form:

Drag force $= a + bv + cv^2$

Where v is the air speed and a, b and c are constants for that particular machine. These characteristics can then be used to predict performance and work out the required gearing, for new power outputs. Note that the drag force here is not entirely aerodynamic drag. It also includes rolling drag, caused by the tyres, brake drag, friction in the wheel bearings and the drive line. Some more power is also absorbed in the inertia of the rotating parts. This does not matter where overall calculations for gearing are concerned, but it does make it more difficult to isolate the aerodynamic component.

Real tests offer several options, which are covered in more detail in Chapter 11. They are:

1 Wind tunnel.
2 Maximum speed.
3 Acceleration.
4 Coastdown.

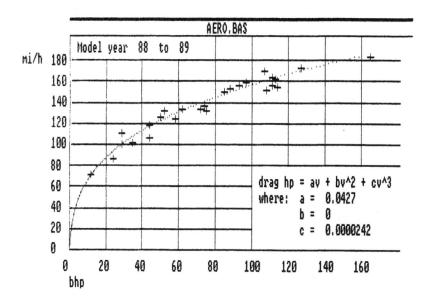

Figure 7.5 A computer plot of twenty-six models tested by *Performance Bikes* in 1988 and 1989, showing maximum speed against peak power (the program for this is given in the Appendix). The program also draws a series of graphs in the form drag hp = $av + bv^2 + cv^3$, where v is velocity and a, b and c are constants which the operator can alter. The curve shown fits the average aerodynamics fairly well and can be used to predict performance quite accurately

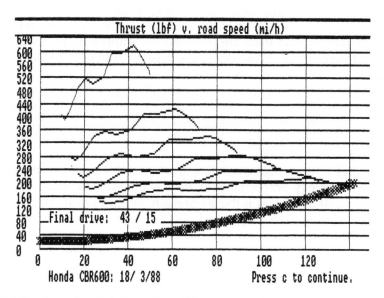

Figure 7.6 The drag values found from Figure 7.5 are repeated here to produce a graph of total drag compared to the road thrust generated by a Honda CBR600. Where the thrust and drag line cross is the maximum speed on that gearing

1 Wind tunnel tests

These tests measure the drag and lift forces on the bike at various speeds and at various yaw angles. The problem, apart from the cost of a full size tunnel, is in compensating for ground movement and wheel movement.

The drag force = $0.5 \, dv^2 \, C_d A$ and as density d and velocity v are easy to measure while frontal area A is not so easy, the value $C_d A$ is often used as a comparator. It tells you the overall slipperiness of the bike but it does not tell you whether a good figure is good because of the aerodynamics or because it simply has a small frontal area.

The German magazine *Motorrad* published the results of tests made in BMW's wind tunnel, in their 21 March 1987 issue, which are summarized in Table 7.1, along with some values for cars obtained in VW's climatic tunnel. Both use $C_d A$ values, so the cars show greater total drag, but an idea of their slipperiness can be seen from the fact that the frontal area of a big bike is in the region of 0.55 to 0.63 m^2 (5.9 to 6.8 ft^2) with no rider. The frontal area of a car like the Ford Escort is quoted as 1.8 m^2 (19.4 ft^2) while a larger car like the Renault 25 has an area of 2 m^2 (21.5 ft^2). Although the cars have roughly three times the area, they have only about 1.5 times the drag force, implying that they are twice as slippery as most of the bikes.

2 Maximum speed

If the power output is known, this is a good comparator for aerodynamic qualities because, at high speed, the air drag is the predominant force. There are problems in finding a suitable location. Most race tracks do not have long enough straights. And there are problems in compensating for wind speed. On-board anemometry is one possible solution; waiting for a calm day is the other.

If the aerodynamics are changed enough to permit a higher maximum speed, then the power characteristics and the overall gearing must be taken into account – the bike will not go faster if this would force the engine to over-speed to a region where the power drops below the level needed for that speed.

Changes in the ambient conditions affect both engine power and drag. Air drag is directly proportional to air density. Standard correction factors for engine tests show that if the air pressure increases or if the temperature falls then both engine torque and air drag increase, but drag increases slightly more than engine power. If the air pressure decreases then the air density falls, but engine torque falls slightly further than air drag. When the air temperature rises, air density falls and this time engine torque falls less than air drag. Any increase in humidity shows up as a real increase in air density as far as drag is concerned but as it is an unburnable addition to the air, it does not increase engine torque (actually there is often an improvement in combustion and heat transfer but the increased air drag still wins).

Table 7.1 Comparison of drag values

Machine	C_dA (m^2) Rider prone	Rider upright
Honda RS500 (1984)	0.243	–
Yamaha TZ250 (1985)	0.269	0.366
Aprilia AF1	0.291	0.444
Yamaha TZR250	0.296	0.421
Bimota DB1	0.319	0.372
Ducati Paso	0.331	0.459
Ducati 750SS	0.341	0.438
Honda CBR1000F	0.349	0.438
Yamaha FZR1000	0.351	0.404
Kawasaki GPZ1000RX	0.354	0.474
Kawasaki GPZ900R	0.361	0.443
Honda VFR750F	0.366	0.447
Suzuki GSX–R1100	0.398	0.430
Honda VF1000F	0.400	0.455
BMW K100RS	0.402	0.429
Suzuki GSX–R750	0.410	0.455
Suzuki GSX1100EF	0.412	0.444
BMW K75S	0.414	0.439
Yamaha FJ1100	0.433	0.483
BMW R100RS	–	0.435
BMW K100RT	–	0.495
Honda XL600 Transalp	–	0.515
Vincent Black Prince (1953)	–	0.562
Kawasaki KLR600	–	0.565
Kawasaki 1000GTR	–	0.605
Porsche 924	–	0.56–0.59
Renault 25TS	–	0.61–0.63
Peugeot 205GL	–	0.61–0.64
Ford Escort 1.3GL	–	0.71–0.75
Citroen 2CV	–	0.84–0.86

Motorcycle data published in *Motorrad*, March 1987; car data by B. Heil.

3 Acceleration tests

Acceleration tests from a standing start, or through the gears, require skill and consistency on the part of the rider, uniform surface conditions and they are difficult to measure. Aerodynamics is only one of four or five factors which can affect the performance. Top gear roll-on tests, especially at higher speeds, are the most reliable (if it is not possible to reach maximum speed), measuring the time or distance to go from one speed to another. Again, it is essential that there is close to zero wind and that the air density does not change by more than a few per cent.

4 Coastdown tests (see Chapter 11)

If the bike is driven to a speed on a level road and then the drive is disconnected, it will decelerate at a rate which is proportional to the total drag acting on it. This is a combination of air drag, rolling drag, driveline drag and drag in the clutch. Aerodynamic forces are greater at high speed, therefore coastdown tests are more appropriate at high speed and if this involves a gear above second, it will not be possible to shift into neutral to disconnect the drive. To see if clutch drag is significant, make two coast down tests close to peak speed in second gear: in one, shift into neutral, in the other, disengage the clutch. If clutch drag affects the results, it will have to be avoided by adjusting the clutch or by confining the tests to the speeds available in second gear.

By timing the interval between two speeds, the bike's acceleration can be found. The total drag force at this speed is the total mass multiplied by its acceleration (note that the apparent mass of the bike will be greater than its static mass because of the inertia of rotating parts), so you have to be careful how you apply the results of this test (see Chapter 11).

It can be used to find a value for C_d or for C_dA. It is also possible to correct for variations in wind speed and air density but the real value of such tests lies in quick, back-to-back comparisons where no such corrections are necessary.

The tests should show up a change in the bike's aerodynamics. They will also pick up changes in the rolling resistance and rotating inertia and can be used, for example, to show the effect of brake pad drag (see Chapter 11). Usually it is enough to know that there is an improvement in performance, so the times on their own are sufficient but it is important to do as many tests as possible, to reject spurious results (which are usually caused by operator error) and to average the rest or to plot a graph which will show the scatter of the results. This scatter is a measure of the resolution of the test and may be as great as the changes you are trying to find.

Finally, coastdown tests are a good way of checking if anything has changed since the previous performance tests. If you do not get the same coastdown times, then whatever has changed will also affect top speed and acceleration tests.

140

Modifications
There are several areas which may be improved:

1   Reduce frontal area.
2   Reduce the projected area at small angles of yaw.
3   Streamline the front.
4   Streamline the sides and rear.
5   Optimize details:
    (a) External shape, protuberances.
    (b) Internal flow.
6   Lift, instability.
7   Behaviour in cross-winds.

Frontal area
Unless it is possible to achieve very low drag coefficients, this is the most important topic. For an unfaired, or poorly-faired, bike it is essential to reduce the area to the minimum.

There are several ways of doing this, but each will be a trade-off with something else and the importance depends on the speed involved. Air resistance increases exponentially with speed: below 60 mph it is very small; above 130 mph it becomes dominant. In between these extremes it may be more important to have better suspension, better ground clearance, better riding position, etc., all of which tend to make the machine bigger.

First, the whole bike can be lowered, which may affect ground clearance. Smaller wheels and lowered suspension (which uses up some suspension travel) are the easiest ways to do this but it may be necessary to raise the engine in the frame and re-route exhausts, etc. in order to keep cornering clearance. Making the engine narrower will help in two ways: first, by reducing the width of the frontal area; and second, by allowing the engine to be fitted lower for the same amount of cornering clearance.

It is often possible to reduce the width of the whole machine by fitting narrower handlebars, smaller radiators and by reducing the size of anything that sticks out. It may be necessary to narrow the bike's waistline to allow the rider to tuck in more effectively. A cramped riding position, in which either the rider's knees or elbows have to stick out, does not help the aerodynamics.

If the riding position can be altered it may be possible to lower the rider and to chop off bodywork at the top. This usually means lowering the handlebars and the seat (by removing the padding if necessary or by making up a smaller subframe) and then shortening the fairing or screen to suit.

The effective size can be reduced by good streamlining. Compared to a naked machine, a fairing might increase the frontal area by 5 per cent, but if it reduces drag by more than 5 per cent, there is an overall gain.

141

Figure 7.7 The classic teardrop shape is confined mainly to record breakers. This is Ernst Henne's 1936 BMW which reached 272 km/h (169 mph). (*K Wörner*)

Yaw angles

In most real conditions the bike is not ridden in still air, or with a perfect head- or tail-wind. There will be some component of crosswind and the oncoming air will see the bike at a shallow angle. Typically, a crosswind of 20 mph and a bike speed of 100 mph will produce a relative angle of just over 11° to the bike's centreline. Seen from this angle the bike is considerably bigger and much less streamlined.

Reducing drag in this range of angles is important because of the increased drag force and also because the side component needs the bike to be steered into the wind to maintain course, which causes some scrubbing at the tyres and increases the rolling resistance.

The rider becomes more prominent and so a compact, well-tucked-in riding position is important. It helps if all vertical edges have large radii. Looking at the bike from an angle of 5-10° will show any unnecessarily bulbous proportions. The angle of slope of the screen, the way the front of the fairing blends in with the mudguard and the way the seat fits the rider are all possible areas of improvement.

142

Streamlined front

A bluff body is just as effective as a pointed body at speeds below 200 mph except where a flat front increases the frontal area in yaw (see above). Generally, it is better to have a slightly wider fairing which takes the air flow smoothly around the hands and levers, rather than a narrow fairing which makes the air flow break up all round its edges.

The sides of the fairing should be angled and radiused so that air deflected from the front wheel is re-attached to the side of the bike. Any ducts for cooling air should not add to the frontal area. They will be better positioned in the turbulent flow behind the wheel or forks, with a ducted exit to a low pressure region to ensure the desired flow.

Lights, indicators and mirrors should be built in, as far as possible and not allowed to stick out. Air intakes to supply the engine need to be located at a high pressure point – in the region of the number plate or the headlamps – with a well-radiused entry and the minimum size needed to provide the required air flow. If air is able to flow across the entrance to the intake then it will be able to lower the pressure at the air box and rob the engine of power.

The fairing can possibly be restyled to add these features and a mudguard built which will shroud the top 60° arc of the tyre, and deflect air flow away from the engine compartment. Note that the mudguard on a road race machine must cover an arc of not less than 30° and not more than 45° measured from the top of the tyre.

Streamlined sides and tail

There has to be a gap behind the screen, so it is at best a deflector and its width and angle are the important factors. Behind the screen there is a downward swirl of air, which is constant in shape at all speeds and can be felt by moving your hand around the edge of the screen while riding along. This works best on a road bike if the swirl hits the top of the helmet and creates just enough draught across the visor to prevent misting and to remove raindrops. Naturally this is with the rider in the upright position.

For minimum drag, the rider will be prone and the screen needs to deflect the swirl of air in a line which will be picked up and continued by the rider's helmet and his back.

The trailing edge of the fairing will cause the air stream to break away. If it does not then the rider's arms and legs certainly will. Under FIM regulations they must be exposed. However, they can be tucked in and there can be a slightly wider tail fairing which could, conceivably, persuade the air flow to re-attach, before tapering off into a slender tailpiece, which would leave a narrow, efficient wake.

The current style of fairing was called 'dolphin' which is ironic because the measured performance of cetaceans far exceeds their power and the reason is that they are so aquadynamic, partly because of their overall shape, but even this is not efficient enough to explain the performance. A submarine built to this shape and given one dolphin-power would not go as fast as one dolphin. The reason for the difference is thought to be in the skin and in the sub-

Figure 7.8 Side and rear streamlining can make a difference but the advantage is lost whenever there is a side wind

cutaneous layer which allows the skin to ripple, following each potential swirl and eddy. If the body shape could do this and prevent the flow from detaching, or make it re-attach immediately, then it would be very efficient, aerodynamically.

The tapering sides and the seat hump can shroud the rear wheel, take in other protuberances such as exhaust pipes and even house ancillary equipment such as electrics or an oil cooler with its exit ducted to the low pressure region behind the bike.

Fairings are not allowed to extend beyond the rear wheel spindle but mudguards, seats and number-plate carriers are.

Detail optimization

Most of the possible development work will fall into this category and, given a particular layout, it is a matter of applying aerodynamic principles to it. Take each component in turn, see if it can be omitted or reduced in size, radius its edges and streamline its tail.

The things to avoid are sharp edges, sudden changes in section or steps, and anything which adds to the frontal area. Most individual changes will be too small to measure by the test methods open to most of us but the sum total should be worth having.

Internal flow can account for a high proportion of the total drag and should be reduced by deflecting the air flow around the front and by limiting the entrance size of ducts. At the same time, it is worth taking care that hot air from a radiator or oil cooler is not directed over ignition coils or the engine intake air system.

(b)

Figure 7.9 (a), (b) and (c) This... plus this... equals this and about another 7 mph, or a reduction in drag of about 17 per cent

145

(a)

(b)

(c)

(d)

Figure 7.10 (a), (b), (c) and (d) Compare the 1987 **(a)** Elf, **(b)** Honda, **(c)** Lawson's Yamaha and **(d)** Mamola's Yamaha with. . .

146

(a)

(b)

(c)

Figure 7.11 The 1988 (a) Suzuki, (b) Yamaha and (c) Cagiva

Some machines run a pressurized intake system, collecting air at the front of the bike and feeding it into an air box. Others run an air box with its intake under the seat of the bike, have open carburettors or individual filters fitted to the carburettors. With a close-fitting tank and bodywork, and a deep-section frame, the area around the intake is often walled in, forming an effective box. It is obviously not an airtight box but the air to it is supplied around the edges which are quite likely to be in a low pressure region when the bike is travelling at high speed. This can alter both the carburation and the power output.

Cooling ducts have a double function. They must provide enough air to keep the engine or its oil at a reasonable temperature. At the same time they need to create the minimum amount of air drag.

The basic rules are:

1   Turbulent air flow in the radiator or engine fins gives better heat transfer.
2   Air drag diminishes with local air speed, so slow air in the radiator intake creates less drag. The duct at the radiator should therefore be larger than the entry.
3   The air intake should not add to the bike's frontal area and it should not disturb the streamline shape. Its entry should have generous radii.
4   The pressure drop across the radiator should be low.
5   Air speed in the radiator and in the exit should be high – which conflicts with requirement 2, unless the radiator shape can straighten out and speed up the air as it passes through.

The radiator frontal area is proportional to the engine heat flow, which is approximately equal to the useful power output. It is inversely proportional to:

1   Vehicle speed.
2   Air density.
3   Air specific heat at constant pressure.
4   The difference in temperature between the air intake and the coolant at the radiator intake.
5   The radiator air speed.
6   The radiator efficiency, which is defined as $(t_1 - t_2)/(t_1 - t)$ where $t_{1,2}$ are the inlet and outlet air temperatures and t is the coolant inlet temperature.

For testing purposes, the optimum operating temperature needs to be established in engine dyno tests, along with the worst case (in a road bike, prolonged idling after a high speed run). The radiator needs to be able to cope with the worst case (or a separate fan has to be provided). Then the pressure drop across the radiator needs to be measured in real conditions and minimized, or a smaller radiator needs to be used.

148

Lift and instability

It is impossible to predict lift. If the air flow over the top of the bike is at a lower pressure than that underneath, there will be an upward force. The drag force, acting at a centre of pressure some distance above the ground, will cause an overturning couple which has the same effect as lift on the front axle. In both cases, weight will be taken off the front axle and this, some studies show, will lead to instability (or will produce instability at a lower speed.)

If the air flow can produce some downthrust it will reverse these tendencies but the only way to detect positive or negative lift is to measure the front suspension travel while the bike is in motion (see Chapter 11). If the forks extend in proportion to the bike's speed, there is positive lift.

The only way to prevent this is to shape the front of the fairing so that it deflects air upwards, to create negative lift. Possibly, if the lift is caused by high pressure under the headlamp region, it would help to put a spoiler under the front of the fairing nose.

As the drag force itself induces lift at the front of the bike, any reduction in drag will also reduce lift.

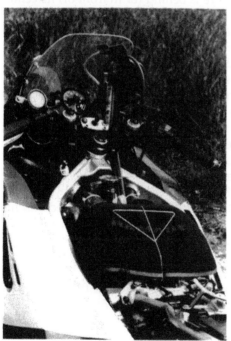

Figure 7.12 To check if the intake air scoops are pressurizing the air box on this CBR600, a manometer has been connected across the air box and the still air region behind the screen

Crosswinds

Because bikes are free to roll and to yaw, and because yaw tends to produce an opposite roll, cross winds can cause quite complicated reactions.

First, a force caused by pressure on the side. This can be demonstrated by using a four-wheel toy with non-steerable wheels. When it is standing

149

still, pushing on its side will only move it when the force is great enough to make the tyres slide (or to make the vehicle roll over if it is tall in comparison to its track). When it is moved along, even a small side force will cause it to travel in that direction – which illustrates the distinction between slip and slide at the tyres. So the force of a sidewind will tend to make any vehicle veer in the same direction as the wind.

Second, the wind will act at the centre of pressure, which will be above ground level, and will produce a torque making the bike roll, also in the same direction as the wind. The size of the roll torque will depend on the wind strength, the side area and the height of the centre of pressure.

Third, any force pushing the bike to one side will, because of steering trail, turn the front wheel to that side. This will initially make the bike steer whichever way the wind pushes it. But (see Chapter 1), steering to the left produces an immediate roll to the right and vice versa. So the presence of the first force will make the bike roll into the wind, countering the effect of the second force. Steeper castor and more trail will increase the effectiveness of this reaction.

Fourth, if the bike's centre of pressure is behind (or in front of) its centre of gravity, then it will behave like a weather vane. The wind force will create a torque making the bike yaw into the wind if the centre of pressure is at the rear, or turn with the wind if the centre of pressure is towards the front. This can add to or subtract from the (third) effect on the steering and set up different slip angles at the front and rear tyres.

In many conditions the effects virtually cancel out one another; bikes can be inherently stable in sidewinds, even in gusting winds. What usually happens is that the bike adopts a slight roll attitude to maintain a straight line and veers slightly with the wind, needing some pressure on one side of the handlebar to hold its original course. Obviously gusting winds buffet the bike, but it responds much faster than the rider can react. The worst thing the rider can do is to try to counter the effect of sidewind gusts; by the time he has responded, the wind force will have changed or disappeared and his reaction will be against a diminished or non-existent force and will only make the bike roll or swerve more than is necessary. He must continue to correct its course, and the best way is to fix a point as far away as possible and steer to keep the bike heading for this point, regardless of the buffeting and roll which the wind causes. He will have to accept that, to hold a straight course, the bike will no longer be vertical and this will restrict its cornering abilities in both directions.

Chapter 8

Frames and chassis detail

The raw requirement for a frame is that it makes a light and rigid platform, holding the wheels in the same plane – or in whatever planes the steering puts them. From the way that suspension has developed this means that the frame has to hold the swing arm bearings and the steering bearings rigidly, in planes which are horizontal for the suspension pivot and vertical for the steering pivot.

These bearings have to cope with engine thrust and with brake forces (in the horizontal direction) and suspension loads (vertically). The frame also has to carry the engine – which may be used to stiffen the frame structure, by becoming the third side of a triangle – and to support the rider and all the ancillary equipment.

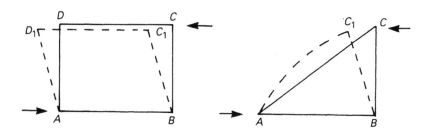

Figure 8.1 A triangle makes a much stronger shape than a four-sided figure because to make it deflect at least one side and all the corners have to deform. A four-sided figure can be deflected if only the corners move.

Historically, frames developed as tubed constructions and there are plenty of examples of light, triangulated 'space' frames which provide a direct and strong connection between the rear wheel and the steering head. Triangulation gives an immense increase in stiffness in the plane of the triangle. If more than three components are joined together to form a structure, i.e., in the shape of a square, rectangle or polygon of more sides, it is easy to deform this shape into a lozenge, diamond or parallelogram because only the joints have to deflect. It can take up the new shape without deforming any of the struts which make up the structure. The stiffness depends entirely on the joints. Stiffness is not always the same as strength. Strength is the resistance to fracture under load, and depends on the material. Stiffness is the resistance to deformation under load and depends on the shape as well as the material.

151

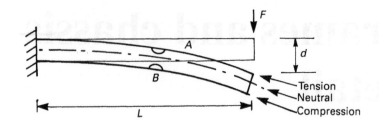

Figure 8.2 When a beam is bent, the convex side is put into tension and the concave side into compression. Somewhere between the two there must be a neutral axis where there is no tensile force. A hole, notch or weld will cause a stress concentration – if it is on the most highly stressed side, e.g. at *A* or *B*, it can seriously weaken the component. The deflection *d* depends on the force *F*, the length and stiffness of the beam, so that $d = FL^3/(3EI)$, where *E* is the modulus of elasticity for the material and *I* is the moment of inertia of the section area

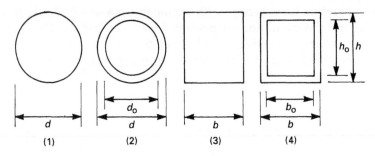

Figure 8.3 (a) Comparisons of the stiffness of different sections:

	Section 1	Section 2	Section 3	Section 4[1]
Section modulus/bending	$\pi d^3/32$	$\pi(d^4 - d_o^4)/32d$	$bh^2/6$	$[bh^3 - b_oh_o^3]/6h$
Section modulus/torsion[2]	$\pi d^3/16$	$\pi(d^4 - d_o^4)/16d$	nb^2h	–
Moment of inertia (axial, to neutral axis)	$\pi d^4/64$	$\pi(d^4 - d_o^4)/64$	$bh^3/12$	$[bh^3 - b_oh_o^3]/12$
Section area (α weight)	$\pi d^2/4$	$\pi(d^2 - d_o^2)/4$	bh	$bh - b_oh_o$

1 The same values apply to an *H, I* or *C* section as long as the section width is the same, i.e., total width is $b - b_o$.

2 The value of *n* depends on the proportions *h:b*

h/b:	1.0	1.5	2.0	3.0	4.0
n	0.208	0.231	0.246	0.267	0.282

If three sides are joined together as a triangle, it can only be deformed if the joints deflect *and* if one or more struts are made to bend, stretch or compress.

As swing-arm suspension developed and bikes were kept narrow, frames became engine cradles which curved around the engine and were quite flexible considering their weight. There were some attempts to make a direct connection between swing arm pivot and steering head, usually with a single, large diameter tube but also with fabricated sheet metal, pressed steel and cast

152

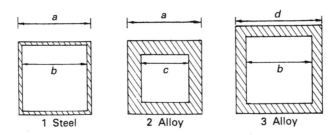

Figure 8.3 (b) Steel has roughly three times the strength and three times the weight of aluminium alloy. A steel box-section tube (1) could be replaced by an alloy component so that (2) the outside dimension and the weight remained the same, or (3) the internal dimension and the weight remained the same. The table below shows the stiffness of the alloy sections compared to the steel section in various wall thicknesses. As alloy has one-third the strength, the stiffness needs to be more than three times greater before there is an overall gain. So when the outside dimension is fixed, steel shows an advantage in stiffness, although when very thin walls are used there is little difference and the extra thickness of aluminium alloy may make it easier to weld, drill, tap etc. When the external dimension is not restricted, alloy shows an advantage in stiffness, for the same weight of material. The same would apply to circular sections. Rectangular, I– or 8– sections and unequal wall thickness can give more stiffness in one direction but less at 90° to this direction. Where strength is unimportant, even higher stiffness/weight ratios can be achieved by using lighter materials such as magnesium alloys, carbon fibre and composites such as aluminium honeycomb.

1 Steel		2 Alloy same weight same dimension a		3 Alloy same weight same dimension b	
a	b	c	stiffness (steel = 1)	d	stiffness (steel = 1)
40	33	8.2	1.86	51.2	4.14
	36	26.2	2.37	47.0	3.63
	37	30.1	2.53	45.4	3.47
	38	33.6	2.69	43.7	3.31
	39	36.9	2.85	41.9	3.15
	39.9	39.7	2.99	40.2	3.02

sections (the so-called monocoque construction). Attempts at fully triangulated space frames were made at regular intervals but their complication and poor access outweighed their potential strength. For the power levels of the old singles and twins, the looped frame was perfectly adequate.

When power levels increased and components like tyres and suspension began a leap-frogging development which let still more power be used, the frame soon became the weak point. By the end of the 1960s Honda had immense trouble with their 500cc GP racer. A few years later, frame development began again with the powerful Formula 750 machines and the unwieldy, 1000cc endurance racers.

(a)

(b)

(c)

(d)

Figure 8.4 (a) The fully triangulated works Norton twin, designed by Peter Williams
(b) Inspired by the purpose-built Godier-Genoud Kawasakis of the early 1970s, endurance teams designed frames literally around their powerplants. This 1974 Honda Suisse machine used a Gold Wing engine.
(c) The 1978 Bimota design had the swing arm pivot concentric with the gearbox sprocket to eliminate changes in chain tension
(d) and (e) The larger section area of light alloys gave a greater stiffness to weight ratio, and sports/racing frames evolved into this beam style, used by Yamaha for the FZR series.

(e)

Figure 8.5 Extra bracing ribs make the section stiffer still. Spondon Engineering used this 8-section in 7020 alloy, which does not need final heat treatment

Developments such as Peter Williams' John Player Nortons – which used space frames and monocoque frames – and the Godier Genoud Kawasaki, led to a succession of new frame designs. As suspension control improved, allowing more suspension travel, it complicated the job of the frame further. As power levels increased, frame stiffness became more important.

Designs slowly changed, from going over the top of the engine to going around its sides (the 'perimeter' or 'A' frame); for example Bimota, Moto Martin, Nico Bakker, Honda RCB. Braced swing arms, to operate suspension linkages appeared, with box section tubing. Where triangulation was not feasible, large box-sections would provide greater strength in bending. Aluminium alloy becomes a strong contender here. At one-third the strength and one-third the weight of steel, alloy has no immediate strength-to-weight advantage and has the disadvantage of less fatigue strength. However, stiffness depends upon the area of material and the distance it is from the axis about which it is twisted (which is why large diameter tubes are stiffer than smaller diameter – but same weight – tubing). Alloy, being less dense than steel, occupies more space for a given weight. So a similar box section, for the same tensile strength and same weight of material, will have a thicker cross-section of material if it is made in alloy and may therefore be stiffer. Conversely, the same stiffness may be had for a reduction in weight. Note that most aluminium alloys – 7020 is an exception – need heat treatment after fabrication.

Where it is possible to get the necessary strength through sheer volume of material, lighter (and weaker) materials have an advantage by giving an increase in stiffness, simply because the section of the part can be larger in area without making it too heavy. So where stiffness and weight are important – wheels for example – magnesium alloy has advantages over aluminium alloy. Carbon fibre has more advantages still, until cost comes into the equation. Composites such as carbon fibre, combined with aluminium honeycomb, can create structures which are both light and immensely stiff in the required directions.

155

Viewed from above, the frame looked like an A. Across the base there was a convenient gap for the suspension strut, air box and electrics. The wide line of the frame made it easy to attach a fairing and the tank sat naturally on the top, low down and with no need for a central tunnel.

The construction of a race frame, even when it follows the same style as a road frame, is very different. Typically, thinner wall sections and lighter, more expensive (and less easily fabricated) materials will be used in competition frames. The result is a surprising loss of weight. The TZ350 and RD350 are very similar in design, yet Table 8.1 shows just how much difference there is.

The potential for frame development is immense – a weight saving in the order of 25 per cent in the case of the Yamaha – and the ability to make radical changes to the frontal area, the weight distribution and the steering geometry. The main consideration for the frame should be the position of the centre of gravity and the approximate length of the wheelbase, along with the maximum stiffness/weight and minimum frontal area of the bike. Details of the steering geometry can be adjusted by altering the ride height(s), by using different yoke offsets or adjustable yokes and by changing the length of the swing arm.

The requirements will differ severely according to the purpose for which the bike is to be used, but the frame can be designed around first the tyres and then the approximate wheelbase and centre of gravity. For a given power output the best layout for traction and for braking can be found (using equations shown in Chapters 5 and 6 or the programs listed in the Appendix). Assumptions about tyre friction have to be based on past tests, on observations of similar machines and on data from the tyre companies. As a first approximation, a coefficient of 0.8 to 1.0 fits the behaviour of sports compound road tyres, going slightly higher, 1.1 to 1.2, for race tyres (1989 season) and higher still for drag racing tyres.

A longer wheelbase will give more stability but slower steering and less weight transfer under braking or acceleration for a given centre of gravity. A spreadsheet computer program is the ideal tool for sorting through the many permutations. Once the wheels and the centre of gravity can be drawn in, the frame then has to accept the engine and rider to keep the centre of grav-

Table 8.1 Weight comparison between Yamaha RD350 and TZ350

Comparison of weight			
Bike	Front axle	Rear axle	Total (lbf)
TZ350	124 (50.7%)	120.5	244.5 No fuel
RD350	155.5 (48.1%)	168	323.5 1/2 tank
	[RD already stripped of bodywork weighing 28 lbf]		

ity in its desired position and then achieve the requirements for:

1 Frontal area
2 Vertical ground clearance
3 Cornering ground clearance
4 Suspension travel

These have different priorities according to the type of bike. Having added the engine and the rider to the drawing, the broad shape of the frame is dictated by these requirements, along with the need for maximum stiffness and minimum weight.

The amount of suspension travel – and the changes which it causes – will decide the layout and type of suspension and there is now enough data to produce a working drawing, from which other details can be sketched in (like changes in driveline, steering geometry, steering lock, turning circle, positions of fuel tank and other ancillaries, etc).

Chassis detail

The design of the bike must take its use into account and be able to accommodate likely crash damage, maintenance and so on. An endurance racer needs quick-change wheels and brake pads; damage-prone parts like footrests need to be easily replaceable. To save fault-finding time, whole electrical circuits need to be mounted on a single, plug-in board.

This may make a difference to the overall design. It may affect things like the choice of wheel size or the type of brakes and so on. Other modifications – such as engine tuning – can alter chassis requirements, for example, by increasing vibration. Careful attention to detail design can make the machine more reliable and easier to use. The reasons for changing things have an order of precedence:

1 Less unsprung weight.
2 Less weight, particularly weight which is furthest from the bike's steer axis or roll axis.
3 To give better protection to the part concerned or make maintenance work on it easier.
4 To improve performance in some other way, such as aerodynamics, rider comfort, etc.

Modifications should be judged against these criteria and the risk of improving one area at the expense of another. It is often easy to make something lighter by making it less reliable. The value of this depends on the length of time the machine has to run.

During the time spent developing and testing the bike it is important to uncover flaws and to establish a 'life' for those parts which wear. Even in short circuit racing, brake pad wear is important because of the need to carry enough brake fluid to replace the worn pads and because thinner pads will heat-up faster and transmit more heat to the fluid. The time taken for tyres to 'go off' or for shock absorber fluid to warm up and lose its damping properties, engine oil consumption, the rate of battery use, are all things which need to be known.

It may be necessary to fit additional instruments during this period, to monitor oil temperature, brake temperature, battery voltage, high and low air pressure regions around the bike, and to keep a constant watch on the tension in various structural bolts and studs.

Frames should be inspected regularly for cracks, for paint flaking off or for the appearance of rust/rusty water on the outside of steel tube. Some frame builders, when developing a new design, have drilled holes in the centre of each welded joint, so every tube communicates with the next tube, have sealed the whole frame, added a Schräeder valve and a take-off point for a pressure gauge or a low pressure warning light. Then the whole frame can be pressurized by a few pounds per square inch and any crack or breakage will show up immediately as a loss in pressure.

Finally, keep a record of the whole thing, including costs. It will help you in the future to plan budgets and make sure that nothing gets overlooked. It is also the sort of thing that impresses future sponsors.

The kind of work that is likely to need doing is listed, alphabetically, below.

1 Brake caliper

If the brake performance has been increased, more heat will be generated and it may be necessary to fit a heat shield behind the pad. Some manufacturers fit them as standard, for example Nissin have a ceramic shield in the four-pot calipers fitted to some Hondas. The alternatives are to use a fluid with a higher boiling point (and which must be changed more frequently) or to provide better cooling. The shield material needs to be incompressible, a poor conductor of heat and able to withstand high temperatures.

Sliding calipers need to be lubricated by using Copaslip on the pins to ensure maximum brake contact. It will also prevent sticking/seizure and stop brake squeal.

2 Brake disc

A small amount of runout is desirable, especially on very high speed machines, because it knocks the pads clear, prevents drag and prevents the pads overheating. Coastdown tests (see Chapter 11) can be used to show how much drag there is. Too much runout will increase lever travel.

Disc overheating will ultimately be shown by the metal discolouring and then cracking. AP market a range of temperature-sensitive stickers and paints, the colour of which indicates the temperature reached and the disc/pad manufacturer will give advice on the maximum permissible temperatures for their products.

3 Brake drums

Brake drums sometimes distort through overheating or wear and the drum goes out-of-round, giving uneven contact and juddering. The drum must be machined concentric or a new liner fitted. The shoes should be set up so that there is even contact when the brake is applied and no drag when it is not.

4 Brake shoes

A leading shoe generates a self-servo force at its leading edge which can make the brake grab and be difficult to control. The effect can be reduced by filing a chamfer on the leading edge, assuming suitable precautions can be taken to avoid any dust produced from the lining, which may contain asbestos. Two- or four-leading shoe brakes must be adjusted so that each shoe makes equal contact with the drum.

5 Brake lines, cables, wiring, etc.

These are all heavy and stiff and can easily restrict steering motion or chafe against the steering head, producing the same symptoms as an over-tight head bearing or too much steering damper. Re-route them so that the steering is not affected. Flexible hydraulic lines should be checked for weakness by looking for bulging while the line is under full operating pressure. There is usually a small ring fitted over the line to use as a gauge. If this will not pass smoothly up the full length of the hose, then the hose should be replaced. (See Chapter 6, Hydraulic hoses and fluid.)

6 Cables

Keep cables thoroughly lubricated (unless they are the nylon type) and check for stretch/adjustment.

7 Chainline

The chain is a source of lost power and rapid wear if it is allowed to run dry or dirty or badly adjusted. If it is too tight it will destroy gearbox bearings and interfere with suspension movement. Once the chainline has been set up accurately, calibrate the adjusters so that future adjustment is quick and reliable. See lubricants, below.

8 Cooling system

On liquid-cooled engines the size and the position of radiators can affect performance (see Chapter 7) and they also add a fair amount of weight. Some can be saved by using the coolant to cool engine oil instead of having separate oil coolers. This can either take the form of a gallery below the radiator for the oil to pass through or a water jacket around the oil filter (some bikes and cars already use one of these methods). Bleed the line to the header tank and use a level plug or disconnect a hose to check that the water jacket and hoses do fill with coolant, with no air locks, after each rebuild.

Coolant will usually be an ethylene glycol-based anti-freeze with corrosion inhibitors for alloy engines, in soft water, distilled water or water that has been boiled. A 25 per cent solution of anti-freeze will give protection down to –13°C; maximum protection is given when 60 per cent anti-freeze is used.

Inspect and clean radiator fins, removing blockages. If fins are damaged over more than 20 per cent of the radiator area then it probably needs replacing.

9 Crankcase breather

This is an important element in engine performance and, because bike engines are so compact, crankcase volume and breathing capacity are usually restricted. Crankcase pumping forms a significant proportion of an engine's power losses. At high revolutions per minute, four-stroke engines throw a lot of oil about and wet-sump engines in particular can blow a lot of oil and oil vapour through their breathers. This is normally trapped in a de-aeration chamber and encouraged to drip back into the engine. Any that manages to get past this trap is fed into the air box to be cycled through the engine and burnt with the fuel. When the air box is removed, as it usually is on competition machines, this breathing/recirculating device goes with it.

To improve engine performance it is necessary to fit the largest capacity breather(s) which can physically fit it. This will relieve crankcase pressure (which also builds up in the cam box) and it will also form a less-restricted path for oil to find its way out of the engine. A catch-tank is obligatory on racing machines; a better solution is a trap which can return oil, based on the OE design, with a catch tank used only as a long-stop. That way you spend less time emptying catch tanks and less money on oil.

10 Electrical equipment

This is likely to be much modified on any tuned machine, either to remove unwanted parts or to add more powerful lighting.

The simplest form is a self-generating ignition unit, which needs no battery and is completely independent. This can be retained if a generator/lighting system is used. The next simplest form is a total loss system, driven by a battery which has to be changed when it runs low.

Putting a generator on the bike makes everything convenient, electrically, but it puts a lot of stress on the engine. Large rotors do not mix very well with high speed engines. If one is used it should, at least, have its taper lapped on to that of the driving shaft and be fitted good and tight. Smaller, competition rotors are available from specialists like Mistral Engineering, and this is a better solution. The manufacturer's race kit may include a suitable alternator if the bike is intended for endurance racing.

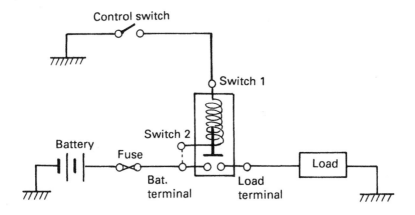

Figure 8.6 Wiring diagram for a relay to operate a heavy electrical load such as twin headlamps

Positive electrical connections encourage corrosion, especially where two or more metals are clamped together and the clamping screw also becomes a convenient electrical connector.

The wire gauge should be selected to suit the electrical load – so twin headlamps require a heavy wiring harness. To keep down the mass of wiring that has to move with the steering, run light gauge wires to the switches and use a relay to supply power to the lights. Wiring instructions and advice on wire gauges should be supplied with the relay. Table 8.2 lists typical connections. Table 8.3 shows wire gauge calculations.

Table 8.2 Electrical relay connections

Connection	Bosch, Cibié, Hella DIN 72 552	Lucas	Marelli	Cartier	Mixo	Others
Battery	30, 30/51	C2	B	5	4	B, BAT, B+
Load	87	C1	H	3	1	
Switch 1	85	W1	P	1	2	S
Switch 2	86	W2	None	2	3	

Battery	=	+12 V supply, full current for load.
Load	=	Connection to load (headlamp, etc.), full current.
Switch 1	=	Solenoid to ground or negative connection, via switch; small current.
Switch 2	=	+12 V supply to solenoid, from battery or from related circuit, may be switched; small current. This may be connected to the '30' (C2 or B) terminal of the relay, as is the case with three terminal relays.

Table 8.3 Wire sizes

Electrical wiring is size coded in the form n/d, where n is the number of strands and d is the diameter of each strand in millimetres, although it used to be in inches. For example 14/.30 means 14 strands, each one 0.30 mm in diameter. The effective cross-section (A) is $\pi d^2 n/4$, so in this case it would be 0.989 mm^2.

To calculate the necessary size for a circuit, first find the current it will take from:

$i = P/v$

Or, $i = v/r$, where i is the current in amps, v is the system voltage, P is the power required by the appliance in watts and r is its resistance in ohms.

The cross section needed is:

$A = iRL/x$

Where R is the resistivity of the material (0.0185 Ω mm^2/m for copper), L is the total length of wire in metres, including any ground connection or return from the appliance and x is the permissible voltage drop in the cable. In 12 V systems x is 0.1 V in circuits taking less than 15 W, going up to 0.3 V for lighting circuits taking more power, and 0.5 V for starter motor leads. Where a relay is fitted, the permitted voltage drop between the switch and the relay is 0.5 V. Halve these values for 6 V systems. Round the cross-section up to the next available wire size (A_1) and then check the 'current density' which is the current divided by the actual cross-section area of the wire (that is i/A_1). For wiring up to 2.5 mm^2, this should not exceed 10 A/mm^2 for continuous operation although it can go as high as 30 A/mm^2 for short bursts, for example horn, starter motor.

Example: for 1 metre of wire to feed two 12 V, 60 W headlamps.
Current $i = 120/12 = 10$ A
Voltage drop in cable, $x = 0.3$ V maximum.
Cable cross-section, $A = 10 \times 0.0185 \times 1/0.3$
$A = 0.617$ mm^2

So the 14/.30 cable would be adequate from this point of view but the current density (10/0.989 or 10.11 A/mm^2) is a bit borderline. For reliability it would be better to go to a heavier gauge, and use a relay so that only light gauge wiring need to be taken to the handlebar switch. A 60 W lamp should be fused at 8 A; two lamps in the same circuit will need a 16 A fuse.

High voltage ignition wiring should not be run close to and parallel with other wiring, including other ignition cables because one can induce currents in the other which may cause spurious sparks or reduce the energy available in the proper spark.

Batteries are heavy and bulky and can often be removed or replaced with something smaller. Conversely, if the alternator is removed a battery may be

Figure 8.7 A rectifier/regulator unit, with a heat sink, built by Peter Roberts to run a moped generator at 12 V and charge the 1.2 Ah Yuasa battery which was the smallest unit able to supply the current for the solenoid valve in a nitrous oxide system

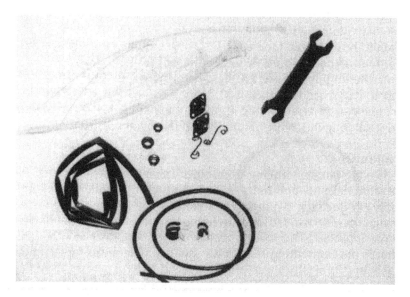

Figure 8.8 Goodridge UK supply a wide range of chassis fittings such as cable clamps, cable ties, shrink-on sleeves for wiring, protective outer sleeves for brake hose, Dzus fasteners, flanged nuts, spanners for brake hose unions, etc.

163

used as a total loss system. The capacity is usually quoted in ampère-hours (Ah) over a ten-hour discharge period. To work out the necessary capacity, add up the current consumption of all circuits in permanent use, ignore intermittent use of indicators, horn, etc. and subtract the current which the generator is able to supply – at the engine speed which will be used. The battery will have to supply the difference and, if this is 2 A, then a 10 Ah battery will last for a maximum of five hours, less if it is not in perfect condition. Often the battery is only needed as a reservoir to cope with short-term, large loads, to act as a stabilizer for things like indicator circuits, to maintain power when the engine is idling and to drive the starter motor. On a competition bike most of this is superfluous and the battery can be replaced by a much smaller unit – particularly one of the compact, jelly-filled batteries or by a capacitor. In this case, a Zener diode-controlled bypass will probably be necessary to take surplus generator power and avoid overcharging the battery. The diode will need to be mounted on a heat sink.

The battery can often be replaced by a suitable capacitor such as a Sprague WB 36D 1721 which is rated at 16 V, 23 A and 68000 μF. Its dimensions are a 2.5 inch diameter and a 4.5 inch length.

If a lead-acid battery has to be used, position it as low as possible and take care that the breather is routed so that it cannot deposit acid on the bike (especially under heavy braking, acceleration or if the bike falls on its side) and that alternative breather exits are blanked off. Electrical components and connections should be protected by using a silicone spray, WD40, a build-up of RTV or petroleum jelly (Vaseline), as appropriate.

11 Engine mounts

Mounts have to contain engine power and vibration, are subject to fatigue and should be inspected for cracks. If flexible mounts are exchanged for rigid plates, then expect vibration to become a problem. Frame tubes which support the engine should not be sprung when the mounting bolts are tightened; use spacers and shims to make sure that the connection is solid. Engine plates can also be used to move the engine back and forth in the frame, so altering the centre of gravity and the weight distribution.

12 Fasteners

There is a big variety, ranging from quick-fasteners (such as Dzus) through to conventional nuts and bolts. A look through a supplier's catalogue will probably produce the required item. Some are available welded to a large base which can be bonded directly into glass fibre mouldings. On standard machines it is not a bad idea to replace all screws which will be disturbed frequently for something of a better quality. There are replacement kits available and typically these offer cap head (Allen) screws to replace the standard cross-head screws. Cap head screws are case hardened and may be difficult to drill for lock-wire. Also, if they are used in a high torque application, the small area under the head will gall and damage the component so a flat airframe washer which is at least as hard as the screw should be used to distribute the load and prevent galling. Textbooks say that bolts should never be used in shear; in chassis components, they almost always are. Consequently the bolt must be strong enough, must not carry shear loads on the threaded portion, and the plain shank must be a good fit in the component. Bolts in single shear (sprocket mountings, disc mountings, calipers, etc.) should be tight. Bolts in double shear are no more than a pin with a thread at the end to prevent it dropping out; there should be a lock-nut or the nut should be wired.

Figure 8.9 The shim has been ground so that the slotted nut aligns with the hole in the spindle when the nut is fully tightened

Bolts and studs are stronger if the thread is rolled, not cut. Their fatigue strength is also greater if the shank of the bolt is reduced to the root diameter of the thread. This is not feasible where the bolt is in shear, but a stress-relieving groove between thread and shank may help. High tensile bolts in En19, En22 up to En24 steel are available and are normally stamped with their property class number (higher is stronger) or with coded marks on the hexagon head.

The best nuts are flanged nuts (or K nuts) although it may be necessary to get them from specialist suppliers such as Goodridge. A specialist will also stock a variety of nuts; there are several classes of fit, ranging from sloppy to needing a spanner all the way along the thread; there are half nuts (lock nuts, which are tightened up against the main nut, which is held by a spanner, in order to lock it by deforming the thread locally between the the two); nylon stop nuts or nuts with deformed threads; castellated nuts and slotted nuts to take a split pin or R-pin. See locking nuts, below. Other fasteners, types of mounting (such as clevis pins, adjustable shackles, spherical (Rose) joints, etc.) may be useful and can be obtained in unlikely places. Try bearing distributors, marine suppliers, ship's chandlers, aircraft breakers, frame builders, race car shops and specialist suppliers of components like brake hose, oil coolers, etc.

13 Footrest hangers.
The first need is for them to be rigid enough to take twice the rider's weight without flexing. It helps, during development riding, if they can offer some

adjustment for the riding position but once this is settled, they can be whittled down to the smallest, strongest size. If the footrest is likely to touch down or hit anything, folding rests are needed and, if crashes are likely, some thought needs to go into the mounting. It is better to let the plate or foot peg break off rather than damage the frame in a crash, and to keep spare pegs, or complete assemblies, to replace the broken parts. The mounting bolts should be the weakest part of the assembly in this case.

Where foot controls are likely to foul parts of the countryside, in moto-cross or enduro, folding pedals will minimize damage and a small chain or cable linking the pedal shank to the frame under the engine will help prevent the pedal being bent back.

The footrest hangers can often form a protective shield to prevent the rear master cylinder from being damaged, or to run the linkage/cable behind.

14 Frame cracks

Steel frames, unless crashed or subjected to abnormal stress or vibration, are pretty impervious to fatigue. The danger signs are paint flaking off the damaged area or rusty water appearing from within the frame tube. Alloy frames, especially light racing frames, are likely to suffer cracking, particularly around highly stressed points such as the steering head, swing arm pivots, engine mountings and anywhere that has a sharp change in section. Welds are usually very strong, the weakest place is alongside the weld. Frames should be inspected carefully at regular intervals, and after any crashes.

15 Handlebars and levers

As the widest part of the bike, the handlebar is vulnerable to crash and collision damage. In events where this is likely it is worth considering shorter handlebars and certainly to move the throttle and levers in-board so that they will not be damaged. In this case, rounded plugs need to be fitted into the handlebar ends. It may be possible to position the levers so that they do not touch the ground when the bike is on its side, or to tighten the clamps so that the levers will rotate before they bend, and to lock the mounting screws to prevent them loosening any further. Plastic hand protectors are available.

Lightweight alloy handlebars are available which will reduce steering inertia but they will not be as strong in a crash.

Various shapes and lengths of lever are also available. It is important that the front brake lever is not able to flex.

16 Ignition wiring

Protect the wiring and connections by using a silicone spray. Switches, etc., should be waterproofed with RTV or Vaseline. Keep high tension lines as short as possible and do not let them run parallel with other wiring or with the frame tubes. If possible, mount the entire circuit on an unpluggable board, so that it can be replaced quickly. Ignition coils and amplifier circuits can overheat – they need to be in a cool air stream, not enclosed and not in the path of air from a radiator or oil cooler.

17 Instruments

Instruments need to be protected from vibration and weather. Loose rubber mounting cushioned by foam is the best answer. It should be positioned in the rider's line of sight when he is in his adopted position on the straight. The rev counter should be twisted so that the red line is in the 12 o'clock position, other instruments should be positioned so that the needle in the normal operating position is vertical, horizontal or aligns with something else so that if it moves, the rider is more likely to notice it.

18 Locking nuts

There are several ways to lock nuts. In each case the main nut should be tightened to the required torque and then held with a spanner while the locking is performed.
(a) Lock nut (half nut). Usually a nut which is only half as tall as a normal nut, although a normal nut will do. In this case the lock nut is tightened against the main nut which *must* be held firmly while this is done.
(b) Thread cement. There are various grades giving different strengths and they must be applied to clean, dry threads.
(c) Nylon insert and deformed thread. Both of these types should only be used once as this tends to cut a thread in the nylon or in the deformed portion of the nut. A similar effect can be made on ordinary nuts (if the end of the bolt is flush with the nut) by using a centre punch on the line of the thread, or by peening over the last male thread.
(d) Lock washer, a washer which is clamped in position, either on splines or by another screw, with a tab which is bent up against the flat of the hexagon. The tabs should be used only once.
(e) Castellated nut and slotted nut. These are used to take a split pin or an R-pin through a hole in the bolt. In this case the nut must be shimmed so that a slot coincides with the hole when the bolt is at the correct tension.
(f) Lockwire. Usually stainless steel wire, in various gauges, sold expressly for this purpose. It is threaded through a hole in the hexagon of the nut, then twisted tightly using self-gripping pliers (or even purpose-built pliers) until it reaches another drilled nut or a suitable anchorage. One end is put through the anchor point and then the two ends are twisted together for a short distance. The lockwire should run at a tangent to the nut, in a direction so that pulling the wire would tighten the nut.

19 Lubricants

(a) Cables (except nylon lined – check with manufacturer) need complete penetration with a light oil – SAE 10 W to 30 – which will need a cable oiler. Aerosol PTFE lubricants are good, if they can get into the cable.

167

(b) Chains. O-ring chains need lubricant to protect then from corrosion – check that aerosols are safe to use with O-rings. Other types need regular cleaning and lubrication. An EP90 gear oil is good enough but some of the chain greases stay in place longer. If the chain is placed in a tray of melting grease, do not allow the grease to boil. Aerosol lubricants have a solvent which lets the lubricant penetrate into the rollers, then the solvent evaporates, leaving the heavier grease inside. Some are good (Motul, Silkolene, Rock Oil, Filtrate), some are not. For dirt bikes it is a good idea to find a chain lube which dries on the outside of the chain and does not encourage dirt to stick to it.

(c) Chassis bearings. An ordinary lithium-based grease is perfectly adequate. A water-repellent type helps and can be used to plug holes around the frame to keep dirt out, making the bike easier to clean after an event.

(d) Wheel bearings. Use a high melting point grease specifically for wheel bearings.

(e) Exhaust flange studs, brake pins. Use Copaslip (an assembly compound which is heat resistant).

(f) General screw threads. A light engine oil will aid assembly and protect them from corrosion, but check what it does to the torque wrench figures.

(g) Frame. Regular, random spraying with WD40 protects against corrosion and makes the bike easier to clean. Clean the discs afterwards.

20 Steering head bearings

Bearings should be inspected frequently on bikes which are used in bumpy conditions or which spend a fair amount of time with their front wheels off the ground. They should be adjusted carefully after slackening the bottom fork yoke clamps. When new taper roller bearings are fitted, the steering stem should be tightened to push the bearing on to its seat, then slackened and adjusted.

21 Tanks

On competition bikes, tanks will be alloy or plastic and need to be mounted on rubber cushions. Breather pipes should be looped so that fuel or oil cannot leak out even if the bike falls over and ball valves in the breather vent should be checked regularly for sticking or blockage. Hoses should be wired in place and, if a fuel filter is used, the flow rate should be checked with a low head of fuel in the tank. At carburettor level it must be capable of flowing $0.6\,p$ pints per hour for a four-stroke or p pints per hour for a two-stroke, where p is the peak power in bhp. This is a minimum requirement, the line should be able to flow considerably more unless you want the fuel tap to do the job of the main jets.

A tunnel in the floor of the tank often means that when one side is empty there is still fuel in the other side. This should be avoided, if necessary by putting another tap in the tank.

If a tank is being made for the bike then it should be baffled to prevent fuel surging back and forth under braking and acceleration, partly to prevent the shifting fuel making the bike pitch and mainly to prevent fuel swilling into the breather or, as the level gets low, swilling away from the tap. If refuelling from a dump tank is planned then a separate breather is needed, because the fuel can only go in as fast as the air can get out.

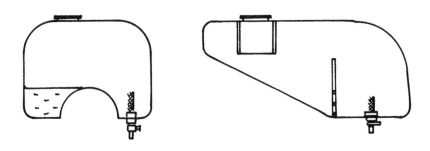

Figure 8.10 Fuel tanks need to be baffled to prevent fuel swilling up into the breather/filler and to prevent fuel swilling away from the tap when the level is low. Avoid tanks with tunnels which can trap fuel in the side opposite the tap, or use two taps

Any new tank needs thorough washing-out and still needs an in-line filter. There is a practice among racing folk of blowing out standard fuel tanks, to increase their capacity without altering their appearance perceptibly, or to make the floor fit the bike exactly to use as a mould for a bespoke tank. It is worth pointing out that this operation is entirely safe if it is done with an hydraulic pump and if the tank is completely filled with water or oil. If it is done with compressed air it becomes a very large bomb.

22 Tightening torque
Because bolts are used to clamp large panels, as adjusters, used in single shear, double shear and, as they were designed, to clamp things together, tightening torques tend to vary a bit. Bolts in single shear need to be tight; others need to be shimmed or spaced out so that there is no free play and if they are not tight then they need some degree of locking.

23 Tyre pressure
Tyre pressures are critical to grip and handling; road/race tyres should not run below 25 psi. Dirt tyres used below this level need clamps to prevent them turning on the rims. The maximum running pressure is usually around 42 psi while the maximum safe overload (to make the tyre seat on the rim) is 50 per cent more than the maximum running pressure, i.e., around 60 psi. Pressures are always quoted for cold tyres but for testing it helps to know

what the pressures are at various tyre temperatures. Race teams often use a cylinder of compressed nitrogen to inflate tyres because the difference in pressure between the cold and fully hot condition is less. Typically, an air-inflated tyre set at 36 psi cold will run at 42 to 43 psi when hot. Measuring pressure is not so easy as most gauges are less than reliable after they have been used for a few months, as a consensus of half a dozen different gauges will show (well, they cannot *all* be right).

24 Vibration

Vibration has two sources: the engine or a wheel which is running out-of-round or out-of-balance. If it is a wheel it will be low frequency and proportional to road speed and can be corrected easily (see Chapter 4).

Engine vibration may be caused by a fault such as a twisted crank or a loose generator rotor or it may be the natural state for that particular engine. Removing balance shafts, fitting heavier pistons, running at higher speed or simply fitting into a different frame can all increase vibration. On singles, twins and any engine which uses the crankshaft webs to partly balance the piston mass, it may be possible to alter the balance factor. This will not reduce the vibration, it will move it into a different plane, to which the frame may not be so sensitive. Fitting flexible mountings or anti-vibration mountings will transmit less movement to the chassis but may cause other problems, for example exhaust fracture, fuel frothing.

Engines with balance shafts, in-line fours, V-motors all have good primary balance but still have secondary out-of-balance which is at high frequency (twice engine speed) but is of a much smaller order than the primary forces. (Only 'boxer' engines, those with six cylinders or more, or

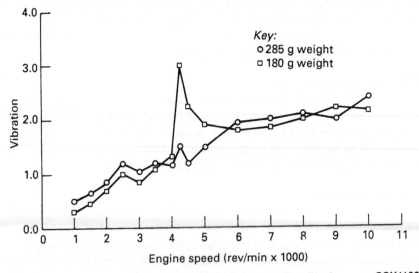

Figure 8.11 Handlebars which went into resonance with engine vibration on a GSX1100E Suzuki. The problem was solved by fitting heavier weights into the handlebar ends. (*Suzuki*)

those with balancers turning at twice engine speed have perfect primary and secondary balance.) Secondary out-of-balance is not usually a problem except when it strikes a natural frequency on some particular part – handlebars are prone to this.

If it only affects one item, the easy solution is to alter the natural frequency of the item (by altering its length), damp out the vibration (by increasing the mass), or isolate it by using a flexible mounting. Most manufacturers put weights in the ends of the handlebars – the illustration shows how dramatic resonance can be, and how effective a few grams can be in changing it.

25 Wheels

Correct any out-of-balance or out-of-round (see Chapter 4). Wheel bearings should be located by a spacer internally and the spindle spaced externally so there is no pre-load on the bearings or on the fork legs (the spindle may need to be tightened before the clamps on the fork legs). Incorrect spacing can lead to brake drag and overheating, sticky fork action and rapid wear in the bushes.

Chapter 9
Bodywork

The bodywork around a bike serves several purposes. The main function is to carry things (fuel, oil, the rider) and it can be extended to protect parts of the machine and the rider from the weather, or in the event of a crash, or simply to isolate them from the vibration and heat of the engine. This line can be extended into aerodynamic bodywork which in turn can be used for styling and to carry number plates (and advertising). Once the bike has been given a hard outer shell it begins to need more complexities, such as ducts to deliver air.

All this adds weight but if the design is economical enough, several jobs can be done with the same piece of bodywork and, despite the weight, it can give an overall gain in performance. Good aerodynamics will improve speed for a given level of power, a good riding position and good ergonomics will help handling, good comfort will reduce rider fatigue which is the same as improving handling over a long period. All that is left is to define 'good'.

Aerodynamics are covered in Chapter 7. Comfort and handling tend to be subjective but this should not mean that it is impossible to be objective about the design. In most cases there will be either a rolling chassis awaiting its seat and controls, or a complete machine on which the seat and controls can be altered, but to a more limited extent. This governs the overall dimensions of length, width and minimum height. Other essentials need to be allowed for at this stage, so check the full suspension travel and the space needed for exhausts, air box, various tanks and electrical components, plus any bulky ancillaries like nitrous oxide bottles and air shifters. At the same time, consider the likely effects of a typical crash – would any of these items cause unnecessary danger or expense; if so, can they be put somewhere else?

By this stage a rough layout will be forming itself more or less by necessity. A scale drawing helps. Once the major components have been placed, the next step is to fit the rider. His weight will be of the same order as that of a competition bike, and it is high up. It will raise the centre of gravity and can move it considerably, backwards or forwards, with serious effects on the handling, braking and rear wheel traction. With a rider in the normal position, his centre of gravity is a few inches in front of his umbilicus, higher or lower depending on how bent his legs are at the hip and knee.

If the approximate centre of gravity of the bike is known (see Chapter 2) and the desired centre of gravity of the entire machine is known (see Chapter 2 and Appendix), then the rider's position can be arranged to suit it. On some machines a variable centre of gravity is desirable, for example on motocrossers and enduro bikes. A rearward centre of gravity helps give more traction on slippery ground. A forward centre of gravity permits more braking force and reduces wheelies when traction is good, so there should be provision for the rider to move about as much as necessary. The higher the rider is, the more significant will his own weight transfer become. Therefore, on a trials bike, the taller the riding position the better.

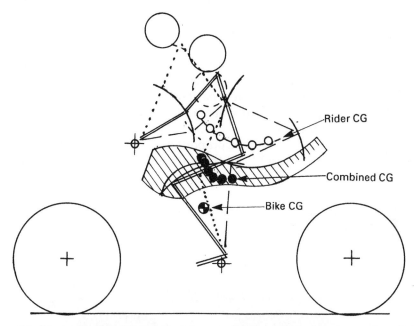

Figure 9.1 Allowing the rider full scope to move can alter the position of the centre of gravity by several inches. On dirt bikes this is critical to maximum traction and will dictate the shape of the seat and the handlebar/footrest position.

The position of the overall centre of gravity can be predicted if the bike's centre of gravity is known. It will lie on a line joining the bike's centre of gravity to the rider's, and will be a distance x from the bike's centre of gravity and y from the rider's centre of gravity so that:

$$x/y = R/W$$

where R is the rider's weight and W is the machine's weight. So putting in the desired centre of gravity for the combined mass of man and machine will dictate where the rider has to be.

Comfort is the next consideration, bearing in mind the need to have the rider move about. On a dirt bike the position of the handlebars and the footrests should allow him maximum movement, forwards and backwards, and the tank, seat, mudguard shape will form an arc traced by the rider's bottom as he moves fully back until his legs are straight, comes forward into the central, naturally sitting position and then extends forward, straightening his legs as he does so.

On a roadster or a roadracer there is less scope for movement but some will be necessary to accommodate the rider's style and to give him the ability to shift his weight sideways for cornering. In general, he will need to move forward and to either side, so the shape of the rear portion of the tank can help. The relation between seat, handlebar and footrest is a critical one: it forms an infinitely variable triangle which has one good set of proportions. The seat is fixed by the centre of gravity requirement. From there, the further the handlebar is stretched out, the further back the footrest needs to be. The reach to the handlebar will depend on the rider's size and on the riding conditions. On a racer he needs to get down as low as possible for speed yet have best visibility and access to controls for cornering. The position of the handlebar then dictates the footrest position, so that the rider can take some weight on the footrests, can move about rapidly and has a natural, comfortable angle at the hip and knee.

A good way to find the right proportions is to sit on a lot of bikes and measure those that feel right *and* those that feel awkward. See where the differences are.

Figure 9.2 The fuel tank has been raised on this RC30 in order to allow more room around the air box.

After the first cut at a riding position it may be necessary to go through the motions again in order to make the footrests higher for ground clearance or to allow for some other feature. When the bike is being tailored to fit the rider, he should wear his normal riding clothing because what feels comfortable in jeans and a shirt can be totally different in leathers and body armour.

The width and angle of the handlebars are the last adjustable factors. Whatever feels best probably is best as long as the rider can get full throttle, reach the brake quickly and control the throttle and brake together. Quick-action throttles are available, although on a four-carburettor layout this can make the throttle action heavy. In practice there will not be too much room for adjustment because of the need to keep 30 mm clearance and a minimum of 20° lock to either side of the centreline (FIM minimum requirements for road racing, along with a minimum handlebar width of 450 mm). For competition machines the permitted dimensions are shown in the *ACU Handbook* which every competitor receives when his competition licence is renewed.

The shape of the bike can now be filled in around the rider, keeping to aerodynamic principles and making each component do as many jobs as possible. For example the front bracket which carries the fairing can also carry any lamps and the instruments, saving the need for separate mountings and removing all this mass from the steering. The seat and side panels can house catch-tanks, plug-in electrics boards and so on.

The problem is making it. There are four or five material options open to the bike builder:

1 Sheet steel.
2 Sheet alloy.
3 Glass fibre (glass-reinforced plastics).
4 Carbon fibre.
5 Sheet plastic.
6 Vacuum-formed plastic.

Sheet metals (1 and 2) are easy to use at a crude level but are difficult to obtain in small quantities and need special equipment and skill to turn into a presentably finished article.

Glass fibre can be strong and light, requires the minimum of equipment and needs patience rather than acquired skill. The same goes for carbon fibre, which is stronger/lighter, at a cosmetic level. At a structural level it needs special equipment which is very expensive.

Sheet plastics have limited use. Vacuum formed plastics can follow any suitable mould – they are not as strong as glass fibre but they can be very light and neat. A look through the *Yellow Pages* directory should reveal any specialists in your area, often under unlikely sounding headings such as plastic sign manufacturers. The moulds follow the same principles as those outlined for glass reinforced plastics below, but will be limited as to size.

Glass fibre

This is very labour intensive for producing one-off items because it is necessary to make a male mould – an exact replica of the component made in wood, plaster, bodyfiller, etc. – and to use this to make a glass fibre female mould. From the female mould, the final glass fibre part is taken. After this it is relatively easy to produce more identical parts, but a lot of work goes into the production of the first one.

Glass fibre can also be used to modify an existing part, in the same way that holes and dents can be filled with a car body repair kit – which will probably be the best material to use as it will have the right ingredients in the right sort of quantities and specific instructions for their use.

For greater quantities, the materials can be bought separately from specialist suppliers, boat-building suppliers, etc. In essence, there is a resin which takes up the shape of the mould and sets hard. The proportion of resin and hardener, the thoroughness of mixing and the temperature at which it cures are very important. The resin is reinforced by strands of glass, carbon or any other suitable material, either in a random matting, or woven in a more regular pattern. It is essential that the matting is fully pressed into the contours of the mould and that the resin completely impregnates the matting, with no air bubbles. This is much harder than it sounds.

The glass commonly used is E glass (calcium aluminabo rosilicate) which is drawn, in molten form, through holes to produce fibres which are 9 to 15 microns in diameter. It has very high tensile strength and easily penetrates skin, so disposable gloves should be worn when handling it. The fibres can be used in several ways: chopped into short strands, loosely bound into ropes or woven into tapes and mats. Other materials are Kevlar and carbon fibre, both of which are available in woven mats. They are stronger, for a given weight, than glass fibre but are much more expensive and are more difficult to use. Unless strength is critical or it is essential to produce a very thin section, there is no advantage over glass.

Male mould

1   This is an exact replica of the desired object – and can often be made from an existing part, suitably modified.
2   It is built up and shaped using wood, two-part polyurethane foam, body filler, sheet metal, wire mesh, more glass fibre and any other material which is convenient. Note that formica (for flat areas), polythene, polypropylene and silicone rubber do not need release agents (see below) and it is often convenient to incorporate these materials in the mould to produce certain shapes (for example, containers).
3   The mould is carved, filled and finally smoothed to shape. A couple of coats of resin should give a smooth finish, but any blemishes here will be faithfully reproduced through the remaining processes. Note that a parallel-sided box is one of the most difficult shapes to release from the mould. Sharp corners are also difficult to reinforce and tend to be formed in brittle resin which is easily damaged, so use generously radiused edges and corners wherever possible. If the shape is such that it cannot be released by the mould, then it must be divided into two halves, splitting the female mould.
4   The finished male mould is covered in release agent – a wax which is polished into the surface to prevent it sticking to the glass fibre.

Female mould

1 This is a glass-reinforced plastic mould taken from the male and therefore will be 'inside out'.
2 A gel coat is applied over the male mould. This is a thixotropic resin which is not reinforced and which forms a smooth, hard outer skin.
3 The glass (or other) mats should be cut up, using shears, to fit the mould. It is important to make the mat follow the contours, and to give a good overlap between successive mats. The resin is workable for about 20 minutes, so this limits the amount that can be done in one go.
4 When the gel coat is tacky, start 'laying up' the glass mats. Mix the resin and the hardener thoroughly, in exactly the proportions given by the manufacturer. Usually the work needs to be done at room temperature (15 to 20°C). The lower the temperature, the longer it takes to cure, but check to see that the material is suitable for lower temperatures if it is necessary to work outdoors. These materials are often poisonous; avoid skin or eye contact; use barrier cream; wear disposable gloves.
5 To get the full strength of the material it is essential that the resin completely impregnates the glass mat. Use rollers (available in various shapes) to achieve this and to squeeze any air bubbles out of the resin. The female mould needs to be rigid, to prevent the final object distorting, and is usually made 1.5 times as thick as the final object will be. Tricky shapes can be strengthened by bonding a wooden or metal former into the female mould, or by using something like a plastic bottle to give a required shape.
6 This should be left for 1 to 2 weeks, tending to be longer if the temperature is low. The female mould can then be separated by tapping around the outside with a rubber or hide mallet. If it is necessary to trim any edges, this can be done with a Stanley knife after 24 hours and the edges sealed with resin.

Finished item

1 In essence, this is a repeat of the steps needed to make a female mould:
   • Release agent.
   • Gel coat resin, with a pigment if you want it to be coloured.
   • Lay-up resin and glass, also with a pigment if you want the colour to go all the way through the material.
2 Woven mats or tape give more strength and are neater for edges, but it is more difficult to impregnate them thoroughly.
3 Moisture in the air will make the cut ends of strands delaminate, especially Kevlar, so if any edges are trimmed, seal them immediately with resin. Any trimming should be done after 24 hours, or when the glass is rigid enough to be cut using a Stanley knife.
4 While laying up the glass mats, keep a uniform thickness, using extra layers to reinforce weak areas or bond in hooks, fasteners, air-frame washers to strengthen mounting points, etc. Kevlar or carbon fibre can be used to reinforce the material.

5   It can be removed from the mould after a few days, depending on the temperature.

Race fairings tend to be very thin and will eventually crack or craze. If one is used in endurance events or modified for road use, it should be reinforced at its mounting points and stiffened wherever it can flex, simply by bonding glass tape or woven mats to the inside surface.

Carbon fibre

Carbon fibre parts can be made in exactly the same way as glass fibre parts, and are a lot stronger and/or lighter. Or carbon fibre can be used to reinforce glass fibre. For structural parts it can be made stronger still if it is cured at exactly the right temperature/pressure/time cycle and for this some very expensive equipment is needed, namely an autoclave oven which is large enough to hold the parts concerned.

In basic respects the steps are the same as with any mould-formed part but because of the potentially huge increases in structural properties they tend to get more scientific, more critical and more expensive where carbon fibre is concerned.

Typically the part – for example a frame – is designed and then optimized using finite element analysis techniques (computer applications which simulate various types of load on a given structure and give deflections and strain

Figure 9.3 This Hejira frame and swing arm are made in carbon fibre – lighter than alloy and measured as more than 20 times as stiff as a steel frame made for the same engine

Figure 9.4 Carbon fibre wheel made by Kudos

Figure 9.5 Raw material – a sheet of carbon fibre pre-preg

Figure 9.6 Aluminium honeycomb is used in composite materials to add stiffness for practically no increase in weight

180

in the material as results). This is expensive but cheaper than making, testing and redesigning the complete component.

Then a mould (a 'tool') is made and the part produced by building up layers of 'pre-preg' – impregnated sheets of carbon fibre which contain the required resin in a sheet made of strands of carbon held parallel and manufactured to exactly the right thickness. More expertise and/or computer software is needed to get the optimum layering and direction of the fibres, combined usually with aluminium alloy honeycomb for stiffness and alloy inserts at 'hard' points where there will be fittings such as engine mounts. To ensure the lay-up matting is forced exactly into the contours of the tool it is often put into a pressure bag – an airtight diaphragm, either with fluid pressure on top of it, or with the air vacuumed out from under it so that it is forced down on to the exact shape of the mould.

In this condition the whole thing is placed in the autoclave, which takes it through the specified temperature-time cycle so that the resin is cured to give optimum strength. Complex parts are usually made in two halves, as with any casting process, and are bonded together afterwards using the same resin.

Chapter 10

Welding

Colin Taylor

The term welding covers a vast area of knowledge. It is like calling working with wood 'carpentry' and including everything from the making of pallets from rough-hewn timber to inlaying rosewood when cabinet-making.

We often need to have something welded because the part concerned has broken. Alternatively, we reckon a particular bit needs to be strengthened to stop it failing. Which of the many welding processes to use, and why, is one of the most difficult things to decide.

Reviewing the readily available processes reveals:

Oxyacetylene metal welding

By virtue of its great versatility most repair shops and nearly all welding establishments will have at least one good, old-fashioned set of cylinders and a torch.

When competition bike frames were made of high quality, cold finished, steel tube with alloying elements, usually chromium and molybdenum added to improve strength, braze welding (wrongly called bronze welding) was *the* welding technique used. This process had the advantage of not melting the tubes and thus creating all sorts of problems because of the intermetallic compounds which can be formed when these elements, chromium in particular, are all molten together. Furthermore, when correctly applied, the technique allowed a build-up of brass at the joint which gave structural strength and the natural, concave, fillet shape of a braze-welded joint provided a gradual transition for any stress which the joint had to handle. The same joint made by, for example, manual metal arc welding (see below) would reveal a very sudden change in section at the joint and a stress concentration more likely to fail under load, as well as the undesirable chromium carbides which would be formed.

Although undoubtedly superseded by more sophisticated processes, as long as the torch is being wielded by a skilled hand, very effective repairs and strengthening can be achieved by the oxyacetylene process but, in general terms, it is now only used for emergency repairs or where very thin materials are being welded.

Metal arc welding

As with the previous process, most welding shops still use and will continue to use this process. Known also as electric arc, arc or stick welding, the process uses an electric power source to produce heat by means of an arc struck between the workpiece and the consumable electrode. When the

182

electrode is of a similar material to the workpiece (e.g. a steel electrode and a steel workpiece) it is melted by the arc and forms part of the weld pool. The work itself is also melted into the pool and thus a fusion weld is made. Non-fusion welds can also be made by the metal arc process but by far the most common type of metal arc weld is the fusion type.

Local heating in the weld zone is very high by virtue of the high energy employed. Therefore, there is a tendency for local stress concentration where a part has been metal arc welded. With thick-sectioned components this is not generally a problem but when used on thin components problems can occur (see *Materials*, below).

The diameter of the electrode determines the current needed and is itself defined by the depth of weld and the length of the arc. Therefore some skill and coordination are needed to maintain the arc and the feed rate, as well as moving the electrode along the line of the weld.

Gas-shielded metal arc welding (MIG, MAG)

This is probably the most common welding technique now used. Often referred to as MIG (metal inert gas) welding, the process uses a consumable wire electrode which is electrically charged and fed into the weld pool through a gun which supplies a shield of gas to surround the whole weld interface.

The nature of the shield gas used gives the process its correct name. When an inert gas is used (helium or argon for example), the process is called MIG. However, when the shield gas is, in chemical terms, active, the process is called MAG (metal active gas). Selection of the shield gas depends upon the reactive nature of the metal being welded. Nickel, chromium and aluminium all demand an inert gas when welding. Steel, however, can be welded when a compound gas such as carbon dioxide or a commercially produced gas such as Argoshield is used as a shield. When the CO_2 shield gas is heated to the high temperatures experienced at the weld interface it decomposes, producing carbon monoxide and free active oxygen. Even though the shield gas changes its nature while the weld pool is molten, it still prevents weld deterioration by attack from gases present outside the shield zone, nitrogen and water vapour in the surrounding air being the most common. The trick to overcome any contamination problem which might result from having free active oxygen in the weld area is to add elements to the welding electrode material which have a natural affinity for active oxygen. Silicon, aluminium and manganese are the most common additives. Titanium and zirconium are also included.

The advantages which gas-shielded metal arc welding has over the metal arc process are many. The skill level required to produce effective welds is less than with the manual metal arc process. The length of weld which can be produced is virtually infinite by comparison with the metal arc process where the length of electrode controls the length of weld which can be deposited before welding has to cease and a new electrode used. Arguably the most significant advantage comes from the very wide range of current which can be used with the same diameter filler wire.

With metal arc welding the diameter of the electrode controls the current which can be used. A 2 mm diameter electrode would have an ideal operating current of about 75 A. To melt a thicker section of parent metal you would need to use a higher current (more energy being needed), for roughly double the right current you would have to use an electrode of about 3.5 mm diameter, any smaller electrode would not be capable of carrying that current. The gas-shielded metal arc process allows a very wide range of current to be used for one diameter of filler wire. As an example, a filler wire of 0.8 mm diameter could happily cope with current from 60 A to 190A. This is because of a linear relationship between the rate at which filler wire is supplied to the weld pool and the current required for melting, known as the 'burn-off rate'. If you need more current (because you are welding something thick) all you have to do is turn up the control and wind up the wire feed rate until the burn-off rate is correct, hence the procedure seen in every welding shop of seemingly depositing weld on a piece of scrap. All that is happening is that the correct burn-off rate is being selected by the welder for the current level he has chosen.

It is much easier to weld this way than it is to try to maintain the ideal arcing conditions which the metal arc process requires. Because of the linear relationship between wire speed and current, when you have selected the correct wire speed to get the burn-off rate you want, the current level will also be correct.

Production runs of bike frames will in almost all cases be welded by the gas-shielded metal arc process, fabricated aluminium frames are welded with an inert gas shield (argon) and steel frames usually with Argoshield as the shield.

To modify or repair one of these items it would be usual to employ an Argoshield or a CO_2 shield MAG weld for steel frames and an argon shielded TIG (tungsten inert gas) process for aluminium ones (which may need subsequent heat treatment).

Gas-shielded tungsten arc (TIG)
The process, also called argon arc, tungsten inert gas or TIG, has some small similarity to the oxyacetylene welding process in as much as the arc is purely used as a heating source to melt the parts being welded. However, the sophistication of the process and, most important of all, the temperature of the arc, allows very high quality welds to be made.

Basically, a non-consumable tungsten electrode is held in a torch which supplies an inert shield gas to envelop the weld zone. Argon is by far the most common gas employed, although in some countries (particularly the USA), helium is employed, mainly on economical grounds, there being no large natural resources of argon in the USA.

By connecting the workpiece and electrode to either side of a welding generator with a high frequency unit an arc can be struck which, when

184

shielded with inert gas will melt nearly all metals. Arc temperatures of 3000+°C are achieved. Tungsten melts at 3410°C and thus the electrode is virtually non-consumable. Slight melt-off of the electrode can occur, particularly if incorrect welding practices are employed, but normally the welding process is quite like gas welding, where a filler rod is melted into the pool created by the arc. Excellent weld purity can be achieved and repairs to aluminium alloy and magnesium alloy parts are usually made in this way.

Other processes
There are many other welding techniques which have great advantages but which, by nature of their very high plant cost, are rarely seen round the corner at your local welders. As production techniques they are widely applied, though. Electron beam welding (EBW) is an exciting production technique. A heated tungsten cathode is used as a highly concentrated thermal energy source, a focused beam of electrons from the cathode being accelerated toward the joint being welded. Voltages up to 150 kV are used to accelerate the electron beam. Very thin (0.1 mm) and very thick (200 mm) sections can be welded in a single pass, with a very narrow weld area. The process is always carried out in a vacuum chamber. Normally parts being EBW processed are completely finish-machined, heat treated and, to all intents, finished prior to welding. An example, such as welding a drive gear or sprocket on to a crankshaft, is where the components can be machined separately and then welded rather than try to hold them together with a key or spline.

Other specialist processes such as plasma-gas metal arc, explosive welding, friction welding, resistance, projection and spot welding all have specific advantages but are not really practical as modification or repair processes.

Materials
The science of the metals to be joined is as involved as the choice of welding process. It is important to have some insight to a component's structure before deciding whether to weld it at all, and, if it can safely be welded, which process to use and what, if any, post-weld treatments are called for.

In very basic terms, metals are constituted from small individual grains. Depending on a number of factors, the size of these grains can be altered, and the smaller the grains, the stronger the part. It can be seen that it is important to have regard for the effect caused by the inevitable heating adjacent to the weld.

Some materials can be strengthened by the addition of alloying elements which serve to form intermetallic compounds within the metal and render them stronger by producing a finer grain structure than is evident in an equivalent, non-alloyed, piece of the same metal.

For reasons of practicality or cost some metals are not alloyed but components made from these metals have to derive their strength from other factors. With steel items, for example, it is possible to improve the strength by heat treatment. To do this, an exact amount of carbon has to be present in the steel. By heating to a specifically high temperature and then cooling at a controlled rate it is possible to significantly alter the strength, hardness and grain size. Any welding which might be carried out on such a treated part would, of course, affect the heat treatment.

Aluminum is another material which is heat treatable, as long as it contains some specific alloying elements, copper, magnesium and silicon being the most usual. Hard, strong intermetallic compounds can be precipitated within the aluminum by utilizing a specific heating, soaking and cooling process. Once again the granular structure can be seen to be made finer, and stronger, by this treatment. Any welding must take account of the heat-treated condition of the material.

When we consider that in basic terms welding is little more than a local casting process we can see the importance of knowing what condition the material is in prior to welding. When materials are not going to be heat treated to improve their strength it is very possible to achieve much the same effect by mechanically deforming the metal to produce smaller grains and hence stronger material. Thus a piece of non-heat-treatable plain mild steel could be made quite strong by cold forming it. We can see examples of this in parts such as steel petrol tanks. Although they are made of very thin, very deformable, non-alloyed mild steel, after they have been stretched over a series of formers to make them petrol tank shaped, they have remarkable strength and rigidity, all because their granular structure has been made finer and finer by the forming processes.

Aluminium items can be 'strengthened' in much the same way but if items which have been strengthened by a mechanical means such as forging or pressing are subjected to the level of heat in a welding process, then, local to the weld, the strength will be reduced because when heating occurs the grains grow back to their 'natural' size. Welding thus tends to anneal the component locally, which can be undesirable.

Components

Before welding anything a series of questions has to be answered. Most important of all being the ability of the part, when welded, to withstand loads in service. Clearly if a part has broken and welding is being used as a repair process, it is important to discover why it broke. Was it subjected to a freak load? In a crash situation, for instance. Did it fail because its design was faulty? Will welding the part leave other parts weakened because of the local 'softening' which will occur? Obviously all these factors cannot be fully addressed but, with common sense it is possible to make a fairly accurate prediction. It is also clearly not possible to list which parts can and cannot be successfully welded. In every case, if there is any doubt, do not repair the part, replace it.

Furthermore it is very important to be able to decide from just looking at a part and discovering from all available information just how it was made. Is it a forging, pressing, die casting, sand casting, alloy steel part, stainless steel and so on? Experience is really the only guide. It is quite obvious that

Table 10.1 Weld processes

Material type (example)	Oxyacetylene (fusion)	Oxyacetylene (braze-weld)	Manual metal arc	Tungsten inert gas-shielded (TIG) arc	Metal inert gas-shielded (MIG) arc	Metal active gas-shielded arc (MAG/CO_2)	Specialist process
Bright mild steel sheet or strip (expansion chamber)	Good	Good	Possible	Good	Optimum	Possible	
Cold drawn steel tube (frame, handlebar)	Reduces strength	Optimum	No	Reduces strength	No	No	
Electric resistance welded steel tube (exhaust pipe)	Good	Good	Possible	Good	Optimum	Possible	
Hot finished steel plate (exhaust mounting bracket)	Possible	Possible	Good	Possible	Optimum	Possible	
Steel forged parts (con-rod, crankshaft)	No	No	Not really	Not really	Not really	In emergency	Electron beam, or laser
Cast iron (cylinder liner, camshaft)	No	Better than glue	Not ideal	Not really	No	No	Eutectic powder process
Ductile iron casting (swing arm ends, some steering heads)	No	No strength	No	No	No	With special filler wire	
Cold finished aluminium alloy sheet (oil tank, chain guard)	Possible	No	No	Ideal	Ideal	No	
Aluminium alloy extrusion (frame, swing arm)	Possible	No	No	Ideal	Ideal	No	
Aluminium alloy cold forging (control lever blades)	No	No	No	Just possible	Not really	No	
Aluminium alloy hot forging (quality brake pedal, handlebar)	No	No	No	Reduces strength	Not really	No	
Aluminium alloy sand casting (cylinder head)	Just possible	No	No	Ideal	Not really	No	
Aluminium alloy die casting (cylinder barrel, crankcase)	No	No	No	Ideal	Possible	Not really	
Alloy steel bar (wheel spindle)	No	No	No	No	No	Not really	
Magnesium alloy casting (racing wheels, engine covers)	No	No	No	Only way	No	No	
Machine-turned steel parts (bolts, studs, nuts, bosses)	Possible	Possible	Possible	Possible	Possible	Ideal	
Hardened steel parts (gear selector)	No	No	No	No	No	No	Electron beam,
Stainless steel (exhaust cover, battery tray)	No	Not really	Possible	Ideal	Possible	No	Electron beam, laser, plasma
Titanium (bolts, exhausts, special parts)	No	No	No	Ideal	No	No	

a cylinder liner, for example, is cast iron but not as easy to decide whether an aluminium part has been die cast or hot forged to shape. In big, fat, broad terms if a part is subject to load, it is more likely to be forged than cast.

An aluminium frame is an example of high class fabrication technique where the external shape of the frame rails can hide sophisticated, internal

187

strengthening ribs. It is possible to improve the relatively low strength of extruded aluminium sections by mechanically deforming the extrusion after production. A local surface deformation pressed or rolled in can give substantial strength improvement at a stress point.

However, welding carried out at the same point would all but negate the benefit. If you are going to attach a bracket, you should closely examine the point to make sure you are not going to weaken the structure. Applying fillet plates to spread the load over a greater area is attractive but the advantages have to be weighed against the weakening effect which welding anything to the frame might have.

To summarize:

• All welded processes will alter the grain structure of the part being welded.
• The important thing to decide is whether the 'softening' effect which welding will have can be accepted.
• The ability to decide which manufacturing processes have been applied to a part and identification of the material from which it is made are the most important factors concerned with correct welding process selection.

Further reading

1   Bainbridge C. G. and Clarke F. *Oxy-acetylene Welding Repair Manual*. BOC Ltd.
2   BOC Ltd. *Argon Arc Welding*.
3   Engineering Industry Training Board. Booklets 3/13, 3/14, 3/15.
4   Handforth J. R. *Practical Aspects of the Argon Arc Welding of Aluminium Alloys*. Aluminium Development Association.
5   Lincoln Electric Co. *Metals and How to Weld Them*. Machinery Publishing Co.
6   Rossi B. E. *Welding Engineering*. McGraw-Hill.
7   Smith A. A. *Carbon Dioxide Shielded Consumable Electrode Arc Welding*. Welding Institute.
8   Welding Institute. *Brazing, Soldering and Braze Welding*.
9   Wiggin Nickel Alloys Co. *Welding, Soldering and Brazing*. Publication 3367.
10 Woods P. F. *Production Welding*. McGraw-Hill.

Chapter 11

Testing and development

Test riding, whether it is to evaluate a new design or to optimize suspension etc., is not easy. It is made difficult by the way in which things overlap: change one component and three or four other factors change as well. It is not helped by the fact that things are seldom what they seem. I have heard two experienced riders describing the same machine, one immediately after the other. One said that the suspension was too hard, while the other thought that it was too soft.

There are psychological factors, too. We all have preferences – rational or not – and sometimes it is important that the rider does not know what has been changed. The rider also benefits from practice as the tests progress, and then suffers from tiredness. It is necessary to check the repeatability of any tests by going back to the first settings later in the day. Some riders are able to go equally quickly on a wide variety of machinery; they either cannot tell the difference or they do not care about it. There is a lot to be said for employing such a rider to race the machine, but not to develop it.

The only lesson is not to accept things at face value; to confirm any results or opinions from at least two criteria. In general, tests fall into one of three categories:

1   Performance tests.
2   Comparative tests.
3   Tests to optimize settings.

Often it will be necessary to optimize the bike before it is possible to make valid comparative tests. For example, because tyres interact with the suspension it must be re-adjusted when a new type of tyre is fitted. This can make the test process quite lengthy, with the risks that the ambient conditions may change, people suffer fatigue or engines wear out. The service life of a piston in a highly tuned engine may be less than ten hours.

1 Performance tests
This is the simplest type of testing. It is rarely necessary to know the absolute performance; the performance relative to some repeatable standard is adequate for most purposes and this can be measured with quite crude equipment.

(a) Speed measurement. The tachometer or speedometer fitted to the machine is a perfectly good comparator. The error varies between a few per cent low (rare) up to 10 per cent high (common) between different instru-

189

ments. On any one instrument the error is likely to be at a minimum in the centre of the scale, increasing as the needle reaches full deflection. Therefore any tests must use the same instrument, in the same speed range. Most instruments are heavily damped, to prevent the needle fluttering, and this means that they lag behind the bike or engine speed when it is accelerating quickly (when this often cancels out the instrument error, so it becomes more accurate during acceleration). For consistent readings, it is essential to let the instrument settle down at a constant reading for several seconds. A change of tyre size will alter the speedometer reading.

A stopwatch and a measured distance can also be used as long as both the operator and the rider are aware of possible parallax errors and the distances involved are large enough for the observer to be able to get a suitable reading. A typical reaction-time for a predictable event like a rider going past a marker is 0 to 0.1 second; if the observer is consistent it will be the same at both the start and stop markers and therefore will not matter. Some people cannot get the same result more than twice out of six timings and should not be allowed anywhere near a stopwatch. If you see changes in the bike of more than 5 per cent you will be lucky, so the timed event has to be quite long (5 per cent of 2 seconds is only 0.1 second, which is in the same order as the error which the timekeeper may produce).

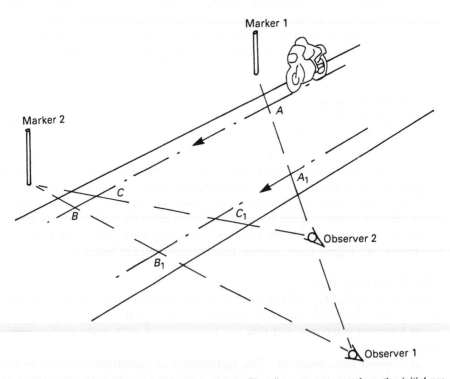

Figure 11.1 Timing a machine between markers. The distance can vary from the initial run (AB) if the rider takes a different line (A_1B_1) or if the observer moves to a different position (AC).

A stopwatch can be used to calibrate the speedometer if you have an accurately measured distance. The police use measured miles and half-miles, with marks on bridges and posts at the roadside, usually with retro-reflective red and blue paint. All normal insurance policies have a clause which excludes use for 'racing pacemaking and speed testing' and I do not know whether calibrating your speedometer at 30 mph could be construed as speed testing.

Going up in sophistication, a pair of light beams to switch a timer on and off is the next most commonly used method. Visible or infra-red light sources and receivers are commercially available. Electronic timers are both common and accurate; all you need is a technician who can couple them together, a suitable length of cable and a surveyor's tape measure. Getting the light beams parallel and accurately measuring the distance are the only difficulties (apart from spurious signals and all the other things which afflict any electrical appliance just when it is least convenient.) The time interval has to be large in relation to the resolution of the timer. We use a gap of 44 yards, because it makes a calculation of $90/t$ to convert a time of t seconds into a speed in miles per hour. A speed of 180 mph gives a time of only 0.5 second, and for the timer to distinguish between 180 and 181 mph, it needs a resolution of 0.00276 second. If, for example, it only read in one-hundredths of a second, it would jump from 176.5 mph to 180 mph and would not be able to distinguish speeds in between. For the same timer the resolution can be improved by making the distance greater – but it then becomes more difficult to measure accurately.

Figure 11.2 Brian Reeson at DataScan, Rushden, built this radar speed-measuring device

Radar is the next speed-measuring option, either commercially-made or custom made. Either way it is expensive, and it has limitations if you cannot get close to the track or if there are a lot of other vehicles.

It has always been theoretically possible to have various sensors on the machine and to record their output – the difficulty has been to make them and the recorder compatible with the conditions. Now that solid state circuits and microprocessors are commonly available, it must be much easier to store the data in a ROM or an EPROM, which can later be read and analysed by a computer.

In fact, data-logging systems of this kind, made by Stack Ltd, and Cranfield Impact Centre are already used on racing cars and a few racing bikes. Typically there are sensors which monitor engine speed, fuel flow, temperature, pressure, suspension travel, throttle position, etc., with an 8– or 16-channel processing unit which can sample the sensors a few times, or many times, every second. The data is stored as binary code, or transmitted via radio telemetry to a trackside receiver. The data is finally manipulated and analysed using a PC type of computer.

The measured speeds are subject to several variables, mainly the wind, but also to slight gradients in the road and bumps on the road surface. The wind speed and direction should be known, several test runs should be made and the amount of scatter noted, the same stretch of track should be used and a series of coastdown tests (see below) should be used to confirm that the conditions have not altered between one set of tests and the next.

Maximum speed is an obvious standard, yet it is open to various interpretations. The gearing has to be optimized, while other factors like rider clothing, rider position, wind speed, etc., have to be kept constant.

One problem is in finding a stretch of track which is long enough. No racetracks in the UK have a straight which will give true top speeds except on bikes which are very small or very slow. Most bikes need a distance of 0.5 to 1.5 miles and room to stop. The tales about speed gradually building up over several miles are not true. It is usually more practical to define maximum speed as the speed reached after accelerating wide open for a

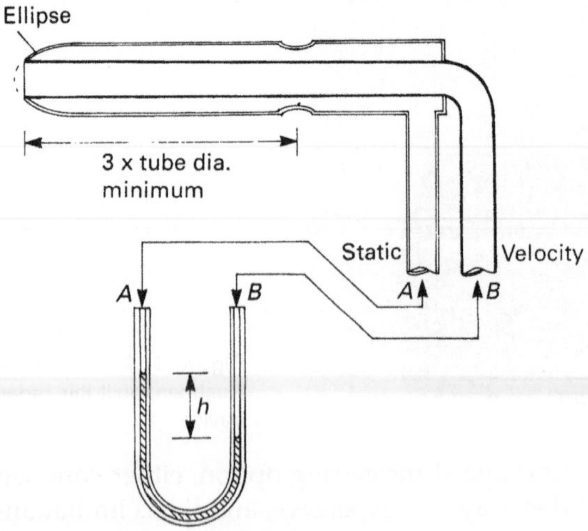

Figure 11.3 Essential dimensions for a Pitot-static tube

known distance – preferably in the same gear, after making an approach at a known speed. This eliminates any error on the part of the rider but it also means that the gearing is critical and must be kept constant – or allowed for, if it forms part of the test (for example, tyre changes).

(b) Air speed. Handheld anemometers are available from ship's chandlers and yacht supply stores. Given the variable nature of wind conditions anyway, this is accurate enough. On board measurements can be taken by making a simple Pitot-static tube.

Figure 11.4 (a), (b) and (c) Pitot-static tube made by Airflow Developments, High Wycombe, along with various manometers and small plastic tapping pieces

The construction details are shown in the drawing. To measure air speed relative to the bike, the tube needs to be mounted well away from the machine, in undisturbed air and to operate at a yaw angle of less than 10°.

The air speed is given by:

$$v = K \sqrt{\{2(p_1 - p_2)/d\}}$$

Where v is the air velocity, $p_{1,2}$ are the pressures at the tip and at right angles to the air flow, d is the air density and K is a constant.

If water is used, the pressure drop $p_1 - p_2$ will be 0.0361 h lbf/in², where h is the head in inches. The expression becomes:

$$v = 12.453 \sqrt{(ht/1.325\,p)}$$

Where v is the air speed in mph, h is the head of pressure in inches of water, t is the temperature in °F$_{abs}$ (°F + 460), and p is the barometric pressure in inches of mercury.

A simple manometer, recording the pressure drop h, can be used to locate regions of high and low pressure around the bike, for instance, to duct air flow to the carburettor intakes or to calculate radiator sizes. Where the direction of the air flow is more important than its quantity, taping tufts of wool in the region will show what is happening, although it may be necessary to have an observer, preferably to photograph the bike.

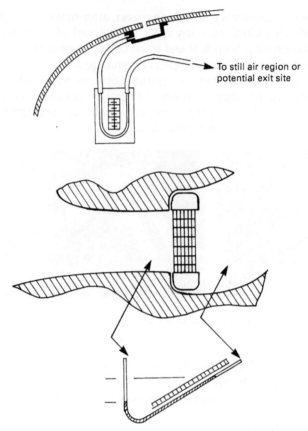

To still air region or potential exit site

Figure 11.5 Manometers can be used around the bike to test for regions of high and low pressure (to site a duct to a radiator, etc.) and to check the pressure drop across radiators so that the optimum size can be used

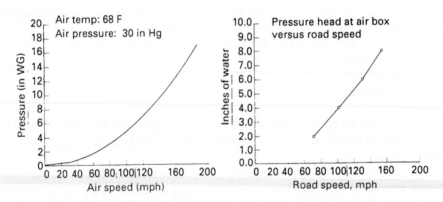

Air temp: 68 F
Air pressure: 30 in Hg

Pressure (in WG)

Air speed (mph)

Pressure head at air box versus road speed

Inches of water

Road speed, mph

Figure 11.6 (a) The relationship between pressure and speed and **(b)** the pressure measured under the tank of a Yamaha OW01. This is not a big enough change to make much difference to the power, but it could have a serious effect on the carburation and would influence the positioning of the carburettor float-bowl vents

194

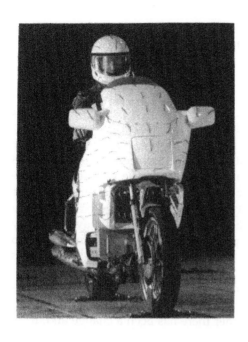

Figure 11.7 Wool tuft tests are used in wind tunnels and in track tests to establish the pattern of air flow in various regions. *BMW)*

(c) Acceleration. This is difficult to measure unless it is possible to have something which can monitor speed continuously. Even then, standing starts require a high level of skill on the part of the rider and constant traction between tyre and road – which cannot be guaranteed. The best way to eliminate these variables is to measure acceleration in a high gear on wide open throttle, and to time the interval between two engine speeds (for example, to begin accelerating at 5000 rpm and to time the interval between 6000 and 8000 rpm) or to measure the speed reached at a certain point (which is often an easier method for a rider on a track). Acceleration tests are subject to the same ambient problems as speed tests.

(d) Coastdown. The bike is taken up to a highish speed and then the drive is disconnected. It is timed between two (or more) speeds as it decelerates. The results of these tests can be used to draw various conclusions about the drag forces acting on the bike and also to establish whether there have been significant changes between one set of tests and another. If the coastdown times change then it indicates that either the bike or the ambient conditions have changed and therefore the test results may not be comparable.

As the machine coasts, it is slowed down by aerodynamic drag, rolling drag, and drag in its driveline, set against its inertia (mass times speed) and the inertia of the rotating parts (which is often regarded as an increase in the bike's apparent mass, just to simplify things). The total drag force is usually expressed as:

Total drag $= A + Bv + Cv^2 + ...$

195

Where v is the velocity and A, B, C are constants. If the deceleration and the mass of the bike are known, then the total drag force can be calculated from:

Drag = mass $\times$ acceleration

Some allowance has to be made for the rotating mass, either by adding it in separately or by assuming that it increases the total effective mass of the bike. To avoid this, the test can be repeated at different speeds and one equation divided into the other. Because deceleration need not be linear, it is usually checked over small speed ranges if it cannot be monitored continuously. From a purely practical point of view it is essential to eliminate other sources of error, for example, same tyre pressures, eliminate brake drag by pushing pads away from disc, use the same part of the track, avoid bumpy stretches, keep the rider in the same position and in the same clothing, keep the same chain tension and the same mass (fuel load). Note that mass is weight divided by g ($= 32.2$ ft/s^2).

If the test is used purely as a comparator (either to tell if ambient conditions have changed or if the bike's drag has been changed) then the times alone will be enough to indicate whether there has been a change and, if so, in which direction.

It is possible to isolate the aerodynamic drag and to produce a characteristic drag curve, which can be used, for example, in programs like RL (see appendix). It is assumed that the drag equation is quadratic, so there are three unknowns: A, B and C. If the total drag is D then

$$D = A + Bv + Cv^2$$

and

$$D = ma$$

Where m is the mass and a is the acceleration between two speeds of which v is the average. If enough tests are done to get equations for three values of a and v, then they can be solved for A, B and C. This is enough to draw the characteristic total drag curve; it includes aerodynamic drag, rolling drag, drag in the driveline and inertia of the rotating parts. As all of these forces have to be overcome by engine power to accelerate the bike, it is not usually necessary to isolate them for practical purposes.

For aerodynamic tests, it is possible to get coastdown results over two speed ranges, v_a to v_b and v_c to v_d. If the first takes time t_1 and the second takes time t_2 then:

Average speed in first test, $v_1 = (v_a + v_b)/2$

in second test, $v_2 = (v_c + v_d)/2$

196

Average acceleration is a_1 and a_2 respectively:

$$a_1 = (v_a - v_b)/t_1$$

$$a_2 = (v_c - v_d)/t_2$$

The drag coefficient, C_d, is given by:

$$C_d = \frac{K(a_1 - a_2)}{A(v_1{}^2 - v_2{}^2)}$$

Where K is a constant which is proportional to the vehicle mass, rotational inertia and air density, and A represents the frontal area of the bike (including rider).

If the test is repeated (using, say, different fairings) at the same speeds, then as long as the mass, air density, etc., do not change, it is not necessary to know the value of K. If the mass changes it would be worth adding weights to compensate. It is also worth arranging the test so that the two comparisons can be made quickly, before there are changes in air density, etc. If the frontal area, A, is changed then this should be accounted for. It is not necessary to know the absolute value, only the change. The most convenient method is to photograph the machine, preferably with a long lens, from exactly the same distance each time, and to have the prints made to exactly the same magnification. Put transparent graph paper over the prints and count the squares covered by the image. If the prints are scaled (by placing a 3-foot rod beside the front wheel, for example) then the area can be found in square inches or feet.

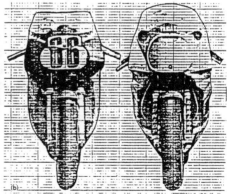

Figure 11.8 One way to compare frontal area is to photograph the machine (and rider), preferably using a lens with a long focal length and to place transparent graph paper over the print, using the grid squares comparatively or scaling them from a reference in the photograph

197

Figure 11.9 (a) and (b) This Leitz equipment measures speed by using a vertical light beam which is focused on the road surface. Speed and time (hence distance and acceleration) are monitored continuously and can be printed out at the end of each test or stored on a disc. The equipment is useful for braking and acceleration tests but its bulk can affect high speed performance and cornering.

If the pictures are not scaled, but the area of the first is equivalent to x squares on the grid and the second is y squares, then the area change is y/x, so the area can be called A in the first test and Ay/x in the second. The tests will then give two drag coefficients (C_{d1} and C_{d2}) in terms of K and A. If one is divided by the other to get C_{d2}/C_{d1}, the terms K and A will disappear, leaving a value such as 0.95, which would mean that the second drag coefficient was 0.95 times the first, i.e., that configuration was 5 per cent more slippery.

There are probably times when it is essential to make this type of test but coastdown tests are limited by the difficulty in measurement and by the changeable nature of the ambient conditions; they are best used to determine whether the conditions *have* changed. The easiest way to see a change in aerodynamics is to use a maximum speed test.

(e) Brake test. The problems with brake tests are that they require a lot of skill from the rider, both in controlling the brakes and in judging the speed. Unless the speed can be monitored continuously, it is difficult to measure immediately before the brakes are used. The bike's kinetic energy is proportional to its speed squared, so small changes in approach speed result in large differences in the stopping distance or time. The point at which the brake is applied is also difficult to measure, and gets harder as the speed increases. Using the stoplight microswitch to start a timer or to fire marker dye on to the road is probably the most reliable method. An alternative is to begin braking at a higher speed and then time the interval between two speeds or over a known distance. Braking at a marker is not reliable at speeds much above 30 mph. Where long distances have to be measured, it is quicker and more accurate to set a datum point in the vicinity of the stop

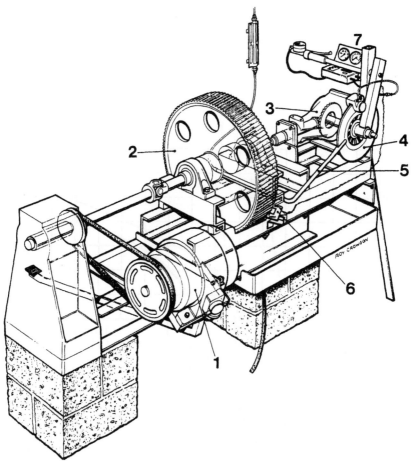

Figure 11.10 Test rig built to evaluate different disc/pad combinations during wet and dry braking. (1) electric motor which is disconnected once the flywheel (2) is up to speed (where its inertia is equivalent to a large touring bike travelling at 30 mph); (3) reduction gear turns the disc (4) at a speed equivalent to 30 mph; (5) nozzle supplies water to the braking surfaces of the disc, via a flowmeter; (6) infra-red light beam broken by teeth on the flywheel, stops an electronic timer when the frequency drops below 25 Hz; (7) console includes hydraulic brake lever, electronic timer activated by the stop-light switch, tachometer and line pressure gauge

area and to measure the distance from there. Data recording devices (see below) can solve most of these problems.

To do any serious brake testing it is necessary to eliminate the problems of rider skill, ambient changes and speed measurement. During the late 1970s, when the wet performance of disc brakes was poor, we built a test rig in conjunction with LEDAR to evaluate different combinations of discs and pad materials. It consisted of two flywheels from Perkins diesel engines (which are stamped with their moments of inertia) which, when driven up to 1500 rpm had roughly the same inertia as a large touring bike travelling at 30 mph. They drove through a Ford back axle, with a disc carrier on a stub shaft instead of the usual half shaft, whose reduction ratio was chosen to turn the disc at the same speed as a bike road wheel at 30 mph. An hydraulic master cylinder and caliper from a Z650 Kawasaki were adapted to fit the rig, with a line pressure gauge and the stoplight switch connected to an electronic timer. A light emitter shone a beam between the teeth of the flywheel to a sensor, with a circuit which counted the passing of the teeth. When the rate of teeth per second dropped to the pre-set value of 25 Hz (12 rpm), it switched the timer off.

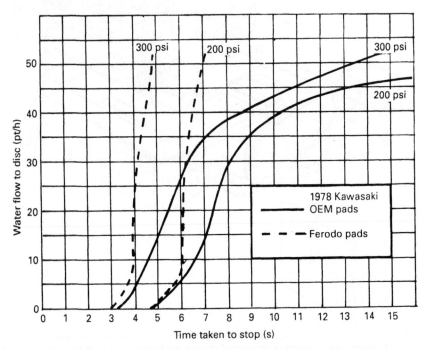

Figure 11.11 Results from the test rig proved that some pad materials were able to function in wet conditions while others were not, at a time (1978) when this was a serious problem on road machines

A nozzle which fitted closely around the disc supplied a variable flow of water via a flowmeter and an electric motor drove the whole thing through a belt drive which could be disconnected. Dry tests at various line pressures showed that the 'stopping' times were in the same order as those for a real bike, and even coastdown times (without the brake) were roughly the same.

The tests showed that, in general, the brake's performance varied in proportion to the water flow until – in certain cases – the flow reached a critical level, beyond which the performance deteriorated badly and became very unpredictable. These cases, which were mainly influenced by pad material, took over 14 seconds to stop, compared to about 5 seconds in dry conditions (at the same line pressure) and about 6 seconds for pad material which was able to work in the wet.

200

Figure 11.12 Control of the water flow proved to be critical. If it was too high, as here, the flow broke up and was splashed away from the disc

Figure 11.13 The nozzle is actually flowing water at a rate of more than 50 pt/h in this shot but the water is sticking to the surface of the disc and travelling with it

201

Data recorders or data loggers can take away most of the guesswork and can make some tests – brake tests for example – very easy to perform. They bring their own problems – mainly in producing so much data that it becomes difficult to handle and to interpret. At a concept level they are very simple to understand. Typically there is a processor with a memory unit whose contents can be downloaded into a computer – usually a portable PC. The unit is connected to various sensors around the bike and is programmed to sample their output at a certain frequency which could be anything from once a second up to several hundred times a second. This data is stored against real time in the unit's memory.

The sensors may be the kind of thing that produces pulses, for example a magnetic speed sensor – a coil which acts like a miniature generator, making an electrical pulse every time a bolt head or the tooth of a sprocket goes past. Others, like suspension or throttle position indicators, are variable resistances which require a voltage to be supplied and the current flowing then determines the resistance and therefore the position of the part which moves. Other sensors record pressure, temperature, fuel flow, take electrical pulses from ignition systems to monitor engine speed, acceleration (from a strain gauge on the cross piece of a small bridge) and so on.

Most of these have an analogue output, which is more susceptible to outside interference and so is converted to digital output and the wiring is carefully screened before being relayed to the memory unit. Other types rely on software to take out the 'spikes' – spurious signals.

Most of the sensors have an arbitrary output – a few volts, a certain range of frequencies; they do not, unfortunately, read in miles per hour or pounds per square inch. So the range of values has to be programmed into the recorder and then the readings have to be calibrated.

Once it is running, the recorder fills up its memory with all the data fed to it and this can later be dumped into a computer, converted into real-life units, plotted as graphs and generally used to analyse the behaviour of the bike.

Having a continous plot of performance is extremely useful but needs a good deal of care in interpretation. Sudden fluctuations in wheel speed could be (a) some spurious electrical signal, (b) surges of wheelspin, (c) increases as the bike banked over and the effective tyre diameter changed or (d) something else. Most results need confirmation from a second source (e.g. engine speed in this case) or from specific tests, such as running the bike at a constant speed, in which case fluctuations would be spurious.

Finally, during any sort of testing, all the fluid levels should be set at known points and their consumption against time or distance should be checked. It gives a baseline comparison for future tests as well as an early-warning of some impending fault.

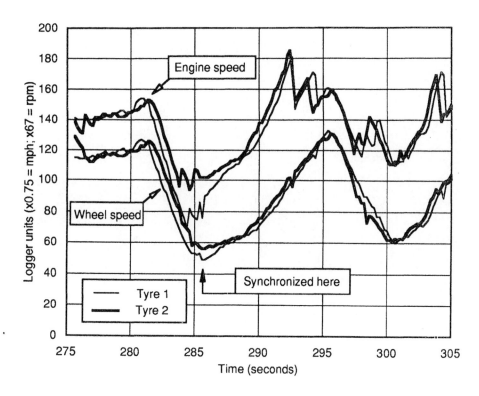

Figure 11.14 Trace from a data recorder used in tyres tests at Oulton Park. The results shown are from sensors monitoring wheel speed and engine speed, using two sets of tyres, along a section of the track approaching the hairpin and accelerating up to the chicane. The values are shown in logger units (i.e. the way it interprets and stores the signals from the sensors) and haven't been scaled to mph or rpm so that they can be shown on the same axis of the graph. It shows how the second set of tyres allowed the bike to be braked later and cornered faster. The traces for the two sets of tyres have been synchronized at the apex of the hairpin. Note that this can give a misleading picture; for example, at the 300 second point, the bike appears to be faster on Tyre 1 than on Tyre 2 ... this is because it is plotted against time and not against distance. The bike was actually faster on set 2 and at the 300-second point it was already in the chicane; on set number 1 it was still braking for the chicane and was actually some distance back down the track, which is why it was travelling at a higher speed at that particular instant. Note also that the rider has arrived at the hairpin one gear higher on set number 1 – they were possibly giving less grip and he took the preceding left-hand curve in a higher gear in order to reduce the torque going through the back tyre. On both sets he let the engine rev out accelerating away from the hairpin (around 292 seconds) and has then short-shifted through the next two gears. This is because there is a right-hand curve on the approach to the chicane where full power couldn't be used and the rider wanted to be in a higher gear when he shut off to start braking for the chicane, to stop the back wheel from hopping or chattering. He begins braking for the chicane at the same point in time on both sets – about 296 seconds – but this means he was further along the track on set number 2, i.e. was able to brake later

203

Figure 11.15 A Kawasaki ZXR 750-J equipped with a data recorder, strapped to the seat hump, and various sensors, one of which can be seen on the swing arm

2 Comparative tests

The brake test rig mentioned above illustrates some of the difficulties of making valid comparisons. It was essential to remove the vagaries of road surface conditions and rider error before the brakes could be analysed closely enough to see what was influencing their performance. Typically, comparisons involve pad material, tyres, shock absorbers, fairings and so on. In most cases the component will change more than one thing and/or the rest of the bike will need to be optimized to get the best performance from the component. So even if the component does not change several things, you will have to. This is not only time consuming, it may also mean that the comparison is not totally valid.

For example, during some tyre tests at Donington (where the transition from power-off to power-on and the subsequent pick-up are more significant than peak power) one set of tyres gave a big improvement in lap times and were giving much higher rev counter readings as the bike went under the bridge on the straight. When these tyres were checked they were much smaller in diameter, although they were the same width as others in the test. Was the improvement owing to better grip or simply because the gearing had been lowered?

Another make of tyre was such a mismatch with the suspension settings that it made the back of the bike bob about and several riders thought the

204

rear damper was fading. So to get a completely fair comparison of all the tyres, it would be necessary to optimize the suspension settings (and probably the tyre pressures) on each, and to alter the gearing to take account of different sizes.

The tests also have to be set up specifically to answer the questions you want to ask. Inevitably there is a lot of subjectivity where tyres, etc. are concerned so it is necessary to ask the riders to do certain tasks and to double check their opinions by measuring speed, lap times and by repeating the same test later in the day. In this case it usually helps if the rider does not know what changes have been made – as long as his ego can stand the possibility of completely contradicting the comments he made earlier.

There is no doubt that some tyres and suspension units do make the bike feel better in ways which are hard to describe, let alone quantify. This is, nevertheless, a valid point in their favour. The converse of this is that if some change makes the bike able to go faster (especially at some tricky part of the circuit) it is more demanding on the rider and he may not feel so happy; give him slightly less grip – or whatever – so that, for the same level of effort the bike arrives more slowly at the tricky part, and he might think it feels better. A speed trap or lap times would disagree.

If it is necessary to spread the tests over several days then it is essential to go back to an earlier setting and if the earlier result cannot be repeated then some sort of correction factor should be applied.

Accepting that some of the testing is subjective, the rider should be asked to note specific points: how much confidence the new set-up gives compared to the last one; how easy it is to turn in to corners or to steer from left to right, whether braking points are later or earlier, where gearshifts have to be made – bearing in mind that it is not easy to remember more than three things at any one time. If a particular point is noted then it may be possible to set up an individual test, for example the revs reached at a certain point may be dictated by the speed allowed through the previous corner. Of course, a data logger will be able to record all of these things – several times every second.

It may be possible to arrange other tests to confirm some particular aspects, for example, during some shock absorber tests the riders noticed that some of the dampers seemed to be fading after a few laps so we arranged for one of the importers to run some tests on a damper dynamometer. Instead of the usual tests, we ran what we thought was the worst damper until there was a significant drop in damping force (which took less than two minutes). Then all of the others were run for over three minutes (by which time the shock absorbers were too hot to be touched) and compared the change in damping force: the riders were right.

Ideally, comparisons need to be made quickly, with the minimum of changes. It is worth arranging the machine so that this can be done, for instance by having the bodywork partially stripped off so that spring struts can be changed more easily. It might also be worth running a quick series of coarse comparisons, without taking the time to optimize the bike or do any fine tuning, so that the rider gets an immediate impression of one unit against the previous one. This would result in a short-list of the best units/ settings which can be tested in finer detail later.

3 Optimizing settings

The difficulty here is in the misleading feedback which suspension can give. Springs which are too soft can cause excessive movement, whose jolts lead the rider to believe that the springs are too hard, especially if the damping has been increased as a sort of compensation. The only certain way to find out is to try alterations in both directions. A deliberate attempt to make any symptoms *worse* is often the fastest way to find what is causing them. It is also essential to keep a note of *all* the settings, then, when everyone is thoroughly confused, it will at least be possible to return to the starting point.

Some of the initial work is theoretical and can be handled beforehand (using something like the computer programs listed in the Appendix). As an example of the steps involved, this was the planned development for a Honda CR500R motocrosser which was to be converted into a hill-climb racer.

It was run on the dynamometer to get power and torque curves, then it was weighed (see Chapter 2) and its centre of gravity found. Hill climbing demands good acceleration from low speed, good braking, handling and grip in slow corners – usually hairpins – and tyres which work without being warmed up, occasionally on mediocre surfaces. The Honda's 13 in of ground clearance and 12 in of suspension travel probably would not be necessary and neither would its motocross tyres.

The tyres are the dominant factor. The softest compound available, and a construction which worked cold, was in a road racing 'full wet'. The large tread pattern causes some squirming with these tyres, and if this proved to be a problem, the plan was to use intermediates or hand-cut slicks with less pattern height. The Honda's forks and swing arm were measured, and there was room to fit the sizes normally used on 250 cc GP bikes. This dictated wheel sizes of 3.50 x 17 front and 4.50 x 17 rear – preferably cast in magnesium alloy because the weight needed to be as low as possible, partly to allow better suspension and partly to reduce rotating inertia and therefore give better acceleration and braking.

Going to 17 in wheels seemed to have a couple of other advantages; it would lower the bike generally and would give the front brake increased leverage.

The figures were fed into the spreadsheet (see Appendix), which showed that the height of the centre of gravity during braking reached a clear optimum at 28 inches with this wheelbase and geometry: compare with Figure 6.19. Any lower and it would lock its front wheel, any higher and it would stand on its nose (see Chapter 6). These limiting conditions could be obtained with moderate hand pressure on the brakes (assuming coefficients of friction in the region of 0.3 for the pads and 0.8 for the tyre) so it looked like the standard brakes would be adequate.

The next step was to feed the power figures into the BASIC program RL, to get an idea of what gearing was needed and what sort of performance figures we could expect. This predicts top speed and acceleration. A few years ago *Performance Bikes* magazine had geared a CR500 for maximum speed and the computer predicted its performance fairly accurately, so good data on the bike's aerodynamic drag was available (it is very similar to a Suzuki GSX- R1100 overall, which suggests that the lighter and slimmer 500 single could be made a little more slippery). The program allows the centre of gravity to be moved around and also calculates whether the limiting condition for acceleration will be wheelspin or a wheelie, as Table 11.1 shows.

Table 11.1 1988 Honda CR500R, predicted performance

	1	2	3	4	5	6	7
Roll radius, in	12.0						
CG height, in	37		30	28[1]			
Tyre, μ	0.8						1.0
Final drive	35/13	35/15			30/15	30/16	
Max. speed, mph	89	103	103	103	120	128	128
Elapsed time, s	6.8	9.0	8.5	8.5	14.1	33.2	33.2
Distance, mi	0.1	0.15	0.14	0.14	0.3	0.67	0.67
0–60 mph, s	4.2	4.2	3.8	3.8	3.8	3.9	3.7
SS$^1/_4$–mi, et/speed([2])	13.2/88	12.5/102	12.2/102	12.2/102	11.9/116	11.9/116	11.9/116
Wheelie[3], mph	0–72	0–72	0–59	0–53	45–55	50–55	50–55
Wheelspin[3], mph	–	–	0–65	0–65	0–65	45–65	–

Notes
1   Optimum height for braking, with this wheelbase and weight distribution.
2   Standing-start quarter mile, elapsed time in seconds and terminal speed in mph.
3   Limiting condition, in this speed range.

The first column shows the bike as stock, but with a rolling radius for a 15/61–17 rear tyre and an assumed coefficient of friction of 0.8. On its stand-ard gearing it would reach only 89 mph and it could wheelie under power up to 72 mph, limiting the power it could put down and restricting its acceleration.

Column 2 shows what happens if the gearing is raised – an improvement in top speed and acceleration, but still restricted by the tall bike's propensity to wheelie. Lowering the centre of gravity, column 3, improves things slightly but it now has the capacity to spin its back wheel as well as lift the front. Lowering the centre of gravity further, to the optimum level for braking, reduces the tendency to wheelie, but only slightly.

Raising the gearing, columns 5 and 6, increases the top speed and also takes the edge off the power, reducing the likelihood of wheelspin and wheelies to an acceptably narrow speed range. Although it has reduced the force at the rear tyre, it has improved acceleration times – partly because it can continue accelerating to higher speeds, partly because the power is not cropped off when the bike tries to wheelie or spin its back wheel.

On a good surface, the racing wet should generate more friction, if the coefficient is 1.0 (column 7) then low speed acceleration will improve.

The last three columns give a target for the final drive gearing and the basic geometry of the bike. At this stage it needs testing, first to check that the assumptions for friction, etc., are realistic, and second to see how much dive and squat there is under braking and acceleration. Putting grease or a cable tie on the fork stanchions will show the suspension travel and with the bike at this ride height the movement of the centre of gravity can be seen. This then needs to be adjusted to get it to the optimum position during braking and acceleration.

There are several ways of lowering or raising the bike – altering the ride height of the suspension, using different length or different rate springs, building in some anti-dive, etc. Some of these could also be used to alter the castor and trail, and the next set of tests would be set up to show whether this was desirable, as well as to confirm that the acceleration times were as predicted and that there was no problem with wheelspin, wheelies or front wheel locking during braking.

If there was, then the data from the tests could be used to revise the computer predictions, changing the value for, say, μ to agree with the measured results. If everything had gone to plan, these tests would concentrate on getting the handling right, particularly for braking down to bottom gear speeds, making a tight turn and accelerating hard. Experiments with castor and trail, weight distribution, damping and spring rates should produce optimum settings - which may not agree with the best settings for traction, in which case it may be necessary to find some other way (for example, to alter the bike's height by using shorter springs instead of the ride height adjusters).

Figure 11.16 (a) and (b) The modified Honda CR500R, fitted with 17-inch Maxton magnesium alloy wheels, Dunlop KR244 soft compound front tyre and KR106 hand-cut slick rear

By this stage the basic layout will be right and any further tests will need to concentrate on detail optimization; finding the best suspension settings, curing problems like wheel patter by altering damper and spring rates, and so on.

The final phase would then be to improve the performance by reducing weight; to investigate the effects of removing one radiator; to alter the riding position (perhaps) for easier cornering or to allow the rider to shift the centre of gravity for optimum acceleration and braking; and to alter the engine performance. Sometimes it is better to change the characteristics of the engine, i.e., the shape of the torque curve, rather than the maximum amount of torque or power that it produces. The program RL can show how such changes would affect traction and acceleration.

During this final testing it is usually difficult to tell what is happening to the suspension, or whether various modifications have affected other parts. Changing the bodywork could alter the air flow to the carburettors, or through a radiator. Some kind of on-board measurements would then be quite useful.

Using old throttle cables (see Figure 11.17) taped against a scale will show the amount of suspension travel, whether the ride height changes (pumping down), how the front and rear match one another and the extent of any wheel patter. This will show the attitude the bike adopts under different conditions and it can then be set up in this position statically to see how the centre of gravity has moved, check the ground clearance, drive line clearance, chain tension and so on. A linear potentiometer (which looks a bit like a steering damper) can monitor suspension travel and feed the results to a data logger to be stored and analysed at a more convenient time.

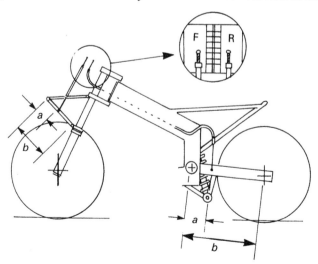

Figure 11.17 A method of checking the ride height while the bike is being ridden. The cables can be mounted on levers to reduce their travel (a/b x wheel travel). This is a useful way to check if suspension is pumping down, allowing wheel patter, or if there is mismatch between front and rear units, etc.

Air flow around the bike can be checked using wool tufts or by making a small manometer to check for high and low pressure areas.

If temperature measurements are needed then a thermocouple can be made, or bought (see Appendix) or temperature-sensitive paint or stickers can be used.

The feel and travel of brake levers can only be assessed subjectively but they can be adjusted to give different amounts of leverage and travel and experiments can be made with different types of hose. A pressure gauge, connected to a T-piece in the brake line, will show the level of force being used and make it easier to calculate the forces operating in the rest of the system.

A manometer connected to a Pitot tube will give air speed readings and can also be used to detect low and high pressure regions around the bike, which will help in producing the most efficient radiator ducts, supplying high pressure air to the engine intakes and so on.

Appendix

Glossary of terms

Acceleration   the rate of change of velocity, in either size or direction; that is, an acceleration is required to follow a curved path, even at a steady speed, because the direction is continuously changing.

Anti-dive   suspension which resists dive (see below), either by increasing the compression damping or the spring force during braking, or by feeding brake torque into the suspension.

Anti-squat   suspension which resists squat (see below) by making the thrust from the driveline try to extend the rear suspension or by making the suspension load-sensitive.

Aspect ratio   a ratio of height/width; e.g. in tyre sections, 180/50 means a nominal width of 180mm with an aspect ratio of 50%, so the tyre section height would be 90mm

Bottoming (out)   when a suspension strut reaches the limit of its travel in compression.

Bounce   one of the types of motion which a vehicle may follow. Bounce is when the whole vehicle moves vertically, staying more or less parallel to the road (compare with *pitch, yaw, roll*).

Bump (of suspension)   compression; the movement of the suspension when the wheel follows a bump.

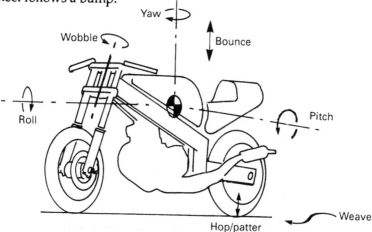

Figure A.1 The axes about which a bike may move

Camber   the angle which a wheel leans from the vertical.

Camber thrust   when a leaning wheel rolls along it tries to roll in an arc, in the direction of the lean, as if it were part of a large cone lying on its side. It would roll in a circle with the tip of the cone at its centre. If it is not allowed to follow this course it sets up a force, or thrust, in the direction which it wants to follow.

Castor   the angle of the steering axis. On bikes with conventional telescopic forks the angle is typically in the range 60–65° measured from the horizontal, or 25–30° measured from the vertical. Both types of measurement are used. Steeper castor creates less camber change when the steering is turned.

Centre of gravity (CG)   the point at which all the mass of a body could be concentrated without altering the balance, etc. of the body. To simplify calculations, all forces such as weight, inertia, centrifugal force, are deemed to act through this point.

Centre of pressure (CP)   when a load is spread over an area (e.g. a tyre contact patch, the area of a brake pad, air resistance acting over the frontal area of a bike) it could be replaced by one single force acting through one point, the centre of pressure, which would have the same effect.

Centrifugal, centripetal acceleration   mean flying away from and accelerating towards the centre of a circle, respectively. When a bike, or anything else, follows a circular course, it has an acceleration towards the centre of the circle, which demands a force in that direction. The inertia of the bike acts in the opposite direction, away from the centre (centrifugal force). Any curved course can be considered to be part of a circle with an instantaneous centre. The acceleration needed to maintain a circular orbit is v^2/r or $r\omega^2$, where r is the radius of the circle and v is the linear speed (e.g. in ft/s) and ω is the angular speed (in $^c/_s$). The force required is the mass of the bike multiplied by the acceleration.

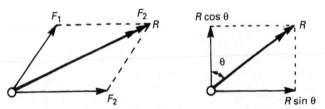

Figure A.2 Two forces (F_1, F_2) can combine to produce a single resultant R. If the length of each line represents the size of its force and the direction is the same as that of the force then the resultant can be found geometrically. Having drawn F_1 to scale and direction (the vector of F_1), draw F_2 also to scale but starting from the end of F_1. The resultant is, in size and direction, the line needed to complete the triangle from the origin to the end of F_2. This may be applied to any number of forces as long as they act through the same point. In the same way, a single force R may be split into two components (usually at 90° to one another) so one will be equivalent to $R\cos\theta$ and the other to $R\sin\theta$)

Chatter rapid suspension or wheel movement, usually on the entry or exit to corners.

Coefficient of drag (C_D or C_d) a number for a given body shape which relates its drag force to its area, air speed and the air density. This number can be used to compare the 'slipperiness' of different shapes.

Coefficient of friction (μ) static: the force required to just move an object across a surface, divided by the weight of the object or the force pushing it against the surface. Dynamic: the force generated parallel to the surface when the object moves, divided by the force pushing it against the surface.

Compression the shortening suspension movement when a wheel is lifted by a bump.

Couple twisting force (torque) caused by two forces acting in opposite directions but not along the same line.

Damping any force which opposes motion or dissipates the energy of a moving part. Specifically, when the suspension is deflected, the energy is stored in a spring which would pass the energy on to the rest of the vehicle; the damping mechanism absorbs some of this energy (usually by forcing oil through small holes); to prevent the spring rebounding with equal force, rebound (or extension) damping is used to take energy away from it.

Dive the tendency of the front suspension to compress during braking (*see Weight transfer*). There are two aspects: the distance that the suspension compresses and the speed with which it does so. Increasing the compression damping will reduce the speed at which the suspension compresses; the distance can be reduced by increasing the spring pre-load, increasing the spring rate or feeding brake torque into the suspension.

Extension (rebound) the travel of the suspension when it is unloaded; the opposite of compression.

Fork offset the forks which carry the front wheel are usually parallel to the steering axis (or nearly so) but a short distance in front of it, in order to provide the desired amount of trail. The distance between the steering axis and the axis of the forks is called the offset. *Note:* the wheel spindle may also be offset from the fork axis.

Frontal area the sectional area of the whole machine projected forwards, or projected in the direction of the air flow over the machine. The area of its silhouette as seen by the oncoming air stream.

g (gravitational constant)   the acceleration of a free body owing to gravity (≈ 9.81 m/s^2 or 32.2 ft/s^2).

Gyroscope   inertia of any rotating mass. Displays certain phenomena, such as precession in which a torque applied to the mass is translated 90° in the direction that the mass is rotating, i.e. trying to turn the handlebar left results in a force which makes the wheel want to lean to the right, whose size depends on the rate at which the applied torque changes.

Hop   rapid, vertical, usually rear, wheel travel, e.g. when too much braking or rapid downshifts make the wheel tend to lock.

Inertia   the tendency of a moving mass to keep moving; it is mass times velocity. In order to change the velocity a force is required (mass times velocity divided by time is equivalent to mass times acceleration, which equates to force). Moment of inertia is a part's inertia about a certain axis and is the sum of all the masses times their distances (squared) from this axis.

Mass   the amount of matter in a body. Not quite the same as weight, although similar-sounding units are used just to confuse us all. Weight is the force exerted by a given mass under the influence of gravity. Force = mass × acceleration and all things accelerate at a rate of '*g*' (see above) under gravity, so weight (lbf) = mass (lbm) × *g*.
When a bike has to be accelerated, braked or turned, it is its mass which has to be accelerated. If a bike weighs 200 lbf then its mass is 200/g or 6.21 lbm.

Moment   torque created by a force about a point (equal to the force times its distance from the point, *see* Torque).

Oversteer   a cornering condition in which (a) the rear tyre's slip angle is greater than that of the front, or (b) an increasing cornering force is produced by a constant or decreasing steer effort. It is a natural tendency for a bike, which is increased by using power. Also called 'drift'.

Pitch   type of movement in which the front of the vehicle moves vertically up or down relative to the rear.

Pre-load   the force with which a part is assembled. In a bearing it is the opposite of clearance and is usually measured as a tightening torque on the nut which holds the bearing (e.g. a steering head or swing arm pivot). Where springs are concerned, it is the fitted load, the load which must be exceeded before the spring will begin to compress. It can be measured as a fitted length (*see* Ride height).

Pro-squat configuration which encourages the rear suspension to compress during braking, usually by feeding some brake torque into the suspension.

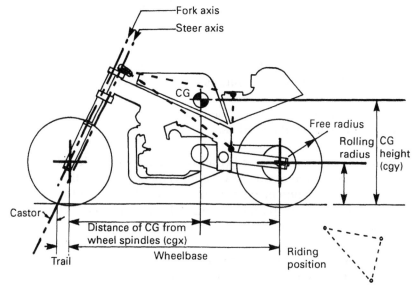

Figure A.3 Some of the dimensions used to describe chassis and steering

Pumping (down/up) characteristic of dampers, when excessive damping force prevents the unit from returning to its neutral position before the next deflection. A series of deflections then produces more travel than the force involved would normally cause.

Rebound same as extension.

Resultant two or more forces (or velocities, accelerations, etc.) can be combined to produce one force (etc.) which is the effective total in both size and direction. See Figure A.2.

Ride height the settled position of the front and rear suspension when a vehicle is in steady motion, or standing still but fully laden. The height of the sprung end of the suspension mounting but, in practice, any convenient vertical measurement between a sprung and unsprung part of the bike,

Rising rate a spring or wheel rate (*see also* Spring rate, Wheel rate) which increases as the suspension is compressed. Air springs do this naturally; coil springs may be progressively wound so that as they compress some coils become coil-bound, thus shortening and stiffening the spring; some linkages reduce the lever ratio as the suspension compresses, giving a rising rate from a linear rate spring.

Roll type of movement in which the vehicle leans to one side.

Roll axis, roll centre the axis about which the vehicle rolls (there is also a yaw axis, pitch axis, etc., but as a bike steers by rolling this mode is the most important).

Shake phenomenon caused between chassis, suspension and tyres which makes the machine judder or vibrate as it approaches the limit of traction.

Skate severe understeer, making the bike run wide in corners because the front will not follow a tighter radius.

Slip, slip angle whenever power is transmitted through a flexible drive, e.g. rubber belt, tyres, there can be relative movement between the driving and the driven parts. The tyre may travel at a speed which is not the same as the bike's speed, yet it is not spinning. It may also travel in a direction which is not the same as the direction in which the wheel is pointing (yet it is not sliding). The slip angle is the angle between the two directions. In these conditions a tyre may generate more frictional force than the normal rolling friction between tyre and road, and certainly more friction than when the tyre is spinning or sliding on the road.

Spring rate the force needed to compress a spring by one unit of length (expressed in lbf/in or N/mm). For a given spring it may be constant (linear), a dual rate (in which the rate increases once the spring has reached a certain deflection) or it may increase progressively.

Sprung (mass) the part of the vehicle which is supported by the suspension, i.e., the frame, engine, bodywork and the rider. *See* Unsprung.

Squat compression of the rear suspension.

Stability a thing is stable if, after being disturbed, it returns to its original position. It is unstable if it does not, or if it takes a long time to do so. Sometimes things return to their initial position so quickly that they build up inertia which carries them past the stable position; if this motion carries on indefinitely, like a pendulum, it is unstable. If the oscillation reduces it is called convergent; if it increases it is called divergent. For practical purposes, if the part has not returned to its start position within a given number of oscillations, say, four or five, then it is considered unstable.

Steer angle the angle at which the wheel, or the handlebar, is turned relative to the centreline of the bike. Note that the handlebar steer angle is not the same as the wheel steer angle, because of the inclination of the steering axis.

Topping (out) when a suspension strut reaches the limit of its travel in extension.

216

Torque turning force, equal to the force multiplied by its distance from the axis of rotation. The distance has to be measured at right-angles to the line of the force.

Trail the distance of the centre of pressure of the tyre's contact patch from the point where the steering axis meets the ground. If the contact patch is in front of the steering axis, then the trail is negative.

Understeer a cornering condition in which (a) the front tyre's slip angle is greater than that at the rear, or (b) increasing steer effort is needed to produce increasing cornering force.

Unsprung (mass) the part of the vehicle which is not supported by the suspension, i.e. the wheels, brakes, moving parts of the suspension, springs and transmission. For the wheels to be able to follow ground contours precisely and for the suspension to be able to control them with minimal shocks being transmitted to the sprung part of the bike, the ratio between sprung and unsprung mass needs to be as high as possible. When a machine is tuned for performance, it is lightened as much as possible and this tends to create problems for the suspension because it is usually easier to lose weight from the frame, bodywork, engine, etc. but not easy to lose a proportionate amount from the wheels and brakes. Therefore the sprung/unsprung ratio tends to become lower.

Vector a quantity, such as speed or force, which has both size and direction. It can be represented graphically by a line whose length is equivalent to the size and which follows the same direction (see Figure A.2).

Weave type of instability in which the bike follows a continuous S-shaped course (as in the warp and weft threads making woven material).

Weight distribution the proportion of the total weight carried by the front and rear axles.

Weight transfer the machine accelerates and brakes via forces generated beween its tyres and the ground, that is the forces are at ground level. The inertia of the bike's mass acts through the centre of gravity, which is some distance above the ground. The distance between the accelerating/braking force and the centre of gravity sets up a couple which tries to overturn the bike (forwards in the case of braking, rearwards during acceleration). Consequently, load is taken off the rear wheel during braking and is carried by the front wheel; during acceleration, the reverse happens. Careful positioning of the centre of gravity can increase the traction available for both acceleration and braking.

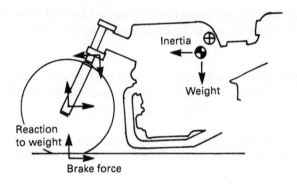

Inertia

Weight

Reaction
to weight

Brake force

Figure A.4 Weight transfer. While the weight tries to rotate the bike clockwise about the front wheel, its inertia during braking tries to rotate it counter-clockwise. The reaction at the front tyre is increased, the increased force compresses the front suspension (which also shifts the centre of gravity slightly and changes the steering geometry)

Wheelbase the distance between the front and rear wheel spindles.

Wheel rate the force required to move a wheel (vertically) through one unit of distance. Measured in lbf/in or N/mm. It is the spring rate multiplied by the leverage between wheel and spring.

Wobble type of instability in which the steering oscillates or 'flutters' rapidly from side to side.

Yaw type of movement in which the vehicle pivots to left or to right without leaning or rolling.

Yoke offset the yokes or triple clamps, carry the forks, at some distance from the steering axis (see fork offset). While the yokes provide offset between forks and steering axis, one may give more or less than the other, in which case the forks will not be parallel to the steering axis. These are known as offset yokes. Some yokes are adjustable.

Symbols used, with typical units

A	area	ft^2, mm^2
a	constant, also acceleration	ft/s^2
b	constant	
c	constant	
C_d	coefficient of drag	
cgx	horizontal co-ordinate of centre of gravity	in, mm
cgy	vertical co-ordinate of centre of gravity	in, mm
C_p, C_v	specific heats of gas at constant pressure and constant volume	$kJ/(kg{-}K)$
cpx, cpy	horizontal, vertical co-ordinates of a centre of pressure	in, mm
d	diameter; also density	in, mm; lb/ft^3, kg/m^3
D	diameter; also drag force	in, mm; lbf
f	final drive reduction	
F	force	lbf, N
g	acceleration due to gravity	$\approx 32.2\ ft/s^2$
G	gear ratio; also shear modulus	N/mm^2
h	height, head of pressure	in, mm
i	electrical current	A
K	constant	
L	length	in, mm
m	mass	lbm
n	number, engine speed in rev/min	min^{-1}
P	force	lbf, N
p	pressure, also primary reduction	lbf/in^2, kPa
R	resistivity	$\Omega\ mm^2/m$
R	force, especially a reaction, also a gear ratio	lbf, N
r	radius	in, mm
r	electrical resistance	Ω (ohm)
S	spring rate	lbf/in, N/mm
t	distance, especially thickness; also temperature and number of teeth on a gear	in, mm; °C, °F
T	torque	lbf–ft, Nm
v	electrical voltage	V
v	velocity	ft/s, mi/h, km/h
V	volume	ft^3, cm^3, litre
w	distance, wheelbase	in, mm
W	weight	lbf, kg
x	distance, especially horizontal axis	in, mm
y	distance, especially vertical axis	in, mm
z	distance, especially at right angles to x and y	in, mm
α	(alpha) angle	° or c
β	(beta) angle	° or c
θ	(theta) angle	° or c
μ	(mu) coefficient of friction	
ϕ	(phi) angle	° or c
ω	(lower case omega), angular velocity	c/s

Computer programs

Program RL: predicting acceleration and top speed

```
10 REM rl
20 REM 12/12/87: 17/4/89
25 REM Maximum speed 220 mi/h
30 REM BASIC2
40 REM jwr
50 REM
70 REM
80 REM
100 GOSUB show: WINDOW #1 TITLE "RL.BAS"
101 REPEAT: rread=INKEY : UNTIL rread>-1
102 IF rread=114 THEN x=30: y=7: GOTO 140
103 CLS
105 INPUT AT(10;6)"Make/model: ",n$: GOSUB nam
106 PRINT: PRINT: PRINT AT(8) "This will be filed as "n1$". Is this
    OK? y/n";
107 REPEAT: nn=INKEY: UNTIL nn>-1: IF nn=110 THEN CLS: PRINT AT(10;
    2)"Full name: "n$: INPUT AT(10;6) "File name (max. 8 characters, no
    space): ",n1$: IF n1$(-3 TO)<>"RL1" THEN n1$=n1$+".RL1"
110 PRINT: INPUT AT(10) "Number of gears: ",y
115 PRINT: PRINT AT(10) "For optimum speed, acceleration and gearsh
    ift"
116 PRINT AT (10) "predictions, give figures for the full engine sp
    eed"
117 PRINT AT (10) "range, up to maximum permissible speed.": PRINT
120 PRINT AT(10) "Number of power/speed entries:"
130 INPUT AT(10) "if less than 20, enter 0  ",x: IF x=0 THEN x=20

140 DIM n(x), hp(x), t(x), ld(x), v(y,x), vr(y,x), f(y,x), g(y): aa
    =0
145 DIM fav(y+1,220), warn(220), a(220), tim(220), thr(220),shift(y
    ), vsh(y)
146 IF rread=114 THEN GOSUB rea: aa=1
148 REPEAT
150 GOSUB menu
160 UNTIL d>=10
200 END

300 LABEL menu
305 WINDOW #1 TITLE "RL.BAS"
310 IF aa=0 THEN GOSUB inpg: GOSUB inpp
315 aa=aa+1
320 CLS
```

```
330 PRINT AT(12;4)"Change gearing data..................1"
340 PRINT TAB(12) "Change power data....................2"
350 PRINT TAB(12) "Display power/torque................3"
360 PRINT TAB(12) "Display thrust/road speed...........4"
365 PRINT TAB(12) "Display acceleration *..............5"
368 PRINT TAB(12) "Display gearshift/traction data 0...6"
370 PRINT TAB(12) "File data *.........................7"
380 PRINT TAB(12) "Read data from file.................8"
390 PRINT TAB(12) "Print data *........................9"
400 PRINT TAB(12) "STOP...............................10"
405 PRINT: PRINT: PRINT TAB(10)"* use item 4 before selecting this"
406 PRINT TAB(10)"0 use items 4 and 5 before selecting this"
407 PRINT: PRINT TAB(10)"Use screen 2 to see current specification.
"
410 INPUT d: ON d GOSUB inpg, inpp, pow, thr, acc,warn, fil, rea, p
ri

430 RETURN


500 LABEL show
505 OPTION DATE 1
510 CLS
520 WINDOW #1 FULL ON

540 WINDOW #1 OPEN
541 ELLIPSE 4150;3000,2500,0.7 WIDTH 5 COLOUR 2
542 PRINT AT(30;6) POINTS(20);"Road Loads"
543 PRINT AT(8;10) ADJUST (16);"Acceleration and terminal speed pre
dictor"

545 PRINT AT(12;18) "Do you want to enter new data"
546 PRINT TAB(12)"or read data from file? n/r "
550 RETURN

600 LABEL inpg
610 CLS: WINDOW #1 TITLE "DRIVELINE SPECIFICATION"
615 IF aa>0 THEN 720
616 d1$=DATE$
620 INPUT AT (10;2) "Primary reduction: ",p
630 FOR i=1 TO y
640 PRINT AT (10) "Internal gear ratio "i;: INPUT ": ",g(i)
650 NEXT
660 INPUT AT (10) "Number of teeth on gearbox sprocket: ",t1
670 INPUT AT (10) "Number of teeth on wheel sprocket;   ",t2
680 PRINT "Do you know the rolling radius of the rear tyre? y/n "
```

221

```
690 REPEAT: e$=INKEY$: UNTIL e$>"":IF e$="n" THEN GOSUB tyre
700 PRINT:INPUT "Rolling radius of tyre (inches): ",r
710 IF aa=0 THEN 860
720 CLS: PRINT AT (10;4) "To change primary reduction press......1"
730 PRINT AT (21;5) "internal gears..............2"
740 PRINT AT (21;6) "gearbox sprocket............3"
750 PRINT AT (21;7) "wheel sprocket..............4"
760 PRINT AT (21;8) "tyre radius.................5"
775 PRINT AT (10;10)"No change..............................6"
780 INPUT bb: ON bb GOTO 790, 800, 810, 820, 830, 870
790 PRINT AT (10;12) "Current primary reduction is "p;: INPUT "New
ratio: ",p: GOTO 860
800 FOR i=1 TO y: PRINT AT (10) "Current gear "i" is "g(i);: INPUT
"New ratio: ",g(i): NEXT: GOTO 860
810 PRINT AT (10) "Current gearbox sprocket is "t1;: INPUT "New spr
ocket: ",t1: GOTO 860
820 PRINT AT (10) "Current wheel sprocket is "t2;: INPUT "New sproc
ket: ",t2: GOTO 860
830 PRINT AT (10) "Current rolling radius is "r: INPUT AT(10) "New
radius: (enter 0 to see tyre sizes) ",r: IF r=0 THEN GOSUB tyre: IF
 r=0 THEN 830 ELSE 860

860 PRINT: PRINT AT (8) "Any other changes? y/n ", :REPEAT: ee$=INK
EY$: UNTIL ee$>"": IF ee$="y" THEN 720
870 GOSUB spec: RETURN

900 LABEL inpp
910 CLS: WINDOW #1 TITLE "ENGINE OUTPUT"
920 PRINT AT (4;2) "Enter speed (rev/min) and engine output (bhp)."

980 PRINT "Enter speed; RETURN; output; RETURN. Enter 1 to stop inp
ut."
990 PRINT: PRINT AT (15) "Rev/min" AT (25) "bhp" AT (35)" torque, 1
b-ft"
1000 w=0
1010 REPEAT
1020 PRINT TAB(15) ">";: INPUT "", n(w);: IF n(w)=1 THEN 1030
1025 PRINT TAB(25)">";:INPUT"",hp(w);: t(w)=hp(w)*5252/(n(w)): PRIN
T TAB(35) ROUND(t(w),1)
1030 w=w+1
1040 UNTIL n(w-1)=1 OR w=x OR hp(w-1)=1
1042 PRINT: PRINT "Are the figures OK? y/n"
1043 REPEAT: e$=INKEY$: UNTIL e$>"": IF LOWER$(e$)="n" THEN CLS: GO
TO 980
1050 RETURN
```

```
1500 LABEL calc
1510 FOR j=1 TO y
1520 q=p*g(j)*t2/t1
1530 FOR i=0 TO w-2
1540 v(j,i)=0.00595*r*n(i)/q: vr(j,i)=ROUND(v(j,i))
1550 f(j,i)=hp(i)*375/v(j,i):       REM vr() is rounded to whole mi/
h
1560 NEXT:                          REM for use in calculating av forc
e
1570 NEXT
1580 RETURN


1600 LABEL thr
1610 GOSUB calc
1620 CLS: WINDOW #1 TITLE "THRUST (lbf) v ROAD SPEED (mi/h)"
1630 PRINT AT(10;4) "Figures or graph? f/g ",: REPEAT: th$=INKEY$:
UNTIL th$>"": IF aa<3 THEN 1640: IF th$="f" THEN 1640
1632 PRINT TAB(10) "Drag factor: option 1,2,3 or 4"
1634 INPUT AT(10) "or select 5 to see drag data: ",ad
1640 IF th$="g" THEN GOSUB gra: GOTO 1740
1650 CLS
1660 PRINT TAB(10) "Road speed" TAB(25)"Thrust"
1665 PRINT TAB(16) "mi/h" TAB(28) "lbf"
1670 PRINT
1680 FOR j=1 TO y
1690 FOR i=0 TO w-2
1700 PRINT USING "          ##.#";,v(j,i),f(j,i): GOSUB wait
1710 NEXT
1720 PRINT: PRINT TAB(8) USING "##.##&"; v(j,i-1)*1000/n(i-1) " mi/
h per 1000 rev/min": PRINT: PRINT
1730 NEXT
1735 PRINT TAB(20)"Press any key to continue."
1736 IF INKEY$="" THEN 1736
1740 RETURN


1741 LABEL wait
1742 IF ww>0 THEN ww=ww+1: GOTO 1770
1743 yy=YPOS
1744 IF yy>750 THEN 1780
1745 PRINT AT(10;20) "Press SPACE bar to continue.";
1750 IF INKEY$<>" " THEN 1750
1760 ww=1: PRINT AT (10;20)"                              "
1770 IF ww=13 THEN ww=0: PRINT: GOTO 1745
1780 RETURN
```

```
1800 LABEL gra
1810 GOSUB drag
1815 WINDOW #1 TITLE "Thrust (lbf) v. road speed (mi/h)"
1820 sx=7000/v(y,w-2): sx=sx/1.05
1830 i=0
1840 j=-1
1850 j=j+1: IF j=w-2 THEN 1880
1860 IF f(1,i)>=f(1,j) THEN 1850
1870 IF f(1,i)<f(1,j) THEN i=i+1: GOTO 1840
1880 sy=4000/f(1,i): i2=i : sy=sy/1.05
1890 k=20: i=0: kk=ROUND(f(1,i)/50): kk=kk*10
1900 REPEAT: LINE (1000+k*i*sx);1000,(1000+k*i*sx);5000: i=i+1
1910 UNTIL k*(i-1)*sx>=5500
1920 FOR j=0 TO i-1: MOVE (800+k*j*sx);700: PRINT j*k: NEXT
1930 i=0
1940 REPEAT: LINE 1000;(1000+kk*i*sy),8000;(1000+kk*i*sy): i=i+1
1950 UNTIL kk*sy*(i)>=4000
1960 FOR j=0 TO i-1: MOVE 400;(875+kk*j*sy): PRINT j*kk: NEXT
1970 FOR i=1 TO y
1980 FOR j=0 TO w-3
1990 LINE  1000+v(i,j)*sx;1000+f(i,j)*sy,1000+v(i,j+1)*sx;1000+f(i,
j+1)*sy
2000 NEXT
2010 NEXT
2020 i=0: REPEAT: dr=a+b*i+c*i^2
2030 PLOT 1000+i*sx;1000+dr*sy MARKER 5 SIZE 1 COLOUR 1
2040 i=i+1: UNTIL dr*sy>3800 OR i>=v(y,w-2)
2045 MOVE 1200;1500: PRINT "Final drive: "t2"/"t1
2046 MOVE 1200;400: PRINT n$": "d1$
2050 MOVE 5500;400: PRINT "Press c to continue."
2060 IF INKEY$<>"c" THEN 2060
2070 RETURN

2100 LABEL drag
2102 IF ad>0 AND ad<4 THEN 2270
2105 WINDOW #1 TITLE "Drag factors."
2110 CLS
2120 PRINT AT(6;2) "Overall drag is assumed to take the form"
2130 PRINT AT(6) "   dr = a + bv + cv²"
2140 PRINT AT(6) " where a,b and c are constants accounting for rol
ling and"
2150 PRINT AT(6) "driveline drag, aerodynamic drag, inertia of the"
2160 PRINT AT(6) "machine and inertia of the rotating parts.""
2170 PRINT
2180 PRINT AT(6) "For large, unfaired bikes (Z1000J, GSX1100)"
```

224

```
2190 PRINT AT(6) "try values of a=16, b=0 and c=0.0105      = opt
ion 1"
2195 PRINT
2200 PRINT AT(6) "For large, faired bikes (GSX-R1100)"
2210 PRINT AT(6) "try values of a=16, b=0 and c=0.0091      = opt
ion 2"
2215 PRINT
2220 PRINT AT(6) "For small, faired bikes (TZR250)"
2230 PRINT AT(6) "try values of a=16, b=0 and c=0.008       = opt
ion 3"
2235 PRINT
2240 PRINT AT(6) "Or new values can be used.                = opt
ion 4"
2250 PRINT
2260 INPUT AT(10) "Option: ",ad
2270 IF ad=1 THEN a=16 : b=0: c=0.0105
2270 IF ad=2 THEN a=16 : b=0: c=0.0091
2280 IF ad=3 THEN a=16 : b=0: c=0.008
2290 IF ad=4 THEN INPUT AT(10)"Value for a: ",a;: INPUT AT(30)"Valu
e for b: ",b;: INPUT AT(50)"Value for c: ",c
2300 CLS: GOSUB spec
2310 RETURN

2400 LABEL acc
2405 IF f(1,1)=0 THEN GOSUB calc: GOSUB drag
2410 GOSUB inpw
2420 GOSUB avg
2430 CLS: t=0: i=0: WINDOW #1 TITLE "Speed (mi/h) v Time (s)"
2440 REPEAT

2470
2480 a(i)=0.682*thr(i)/m
2490 tim(i)=1/a(i): t=t+tim(i): REM t() in seconds
2495 IF i=60 THEN t60=t
2500 i=i+1: UNTIL i-1=v(y,w-2) OR thr(i)<=0: vmax=i-1: etmax=t: j=i
-1
2510 sx=6800/t: sy=4000/(vmax+10): i=0
2515 IF etmax>38 THEN n=4 ELSE n=2
2516 IF etmax>76 THEN n=8
2520 REPEAT
2530 LINE 1000+2*i*sx;1000,1000+2*i*sx;5000
2540 MOVE 800+n*i*sx;700: PRINT n*i
2550 i=i+1
2560 UNTIL n*i*sx>=7000
2570 i=0
```

```
2580 REPEAT
2590 LINE 1000;1000+i*sy,8000;1000+i*sy
2600 MOVE 450;900+i*sy: PRINT i
2610 i=i+20
2620 UNTIL i*sy>=4000
2625 dist=0: et=0: et4=0: d4=0: sp=0
2627 gr=sg
2630 FOR i=0 TO vmax-1

2640 dist=dist+(tim(i)*(2*i+1)/2)/3600: et=et+tim(i)
2650 LINE 1000+et*sx;1000+i*sy,1000+(et+tim(i+1))*sx;1000+(i+1)*sy
2652
2655 IF shift(gr)=i THEN 2656 ELSE 2660
2656 LINE 1000+et*sx;1000+i*sy,1050+et*sx;820+i*sy
2657 gr=gr+1
2660 IF dist>0.23 AND et4=0 THEN GOSUB quart: PRINT d4: PLOT 1000+e
t4*sx;1000+v4*sy MARKER 2
2670 NEXT
2680 dmax=dist: IF dist>0.2494 THEN 2685
2681 REPEAT: dist=dist+(tim(i)*i)/3600: et=et+tim(i):UNTIL dist>0.2
499
2682 et4=et: v4=i-1: REM for bikes which reach max speed inside 1/4
mile

2685 dmax=ROUND(dmax,2):vmax=ROUND(vmax,1):etmax=ROUND(etmax,1):et4
=ROUND(et4,2):v4=ROUND(v4,1):t60=ROUND(t60,2)
2690 MOVE 4500;2000: PRINT "Max speed: "vmax"mi/h in "etmax"s "
2700 MOVE 6500;1750: PRINT "and "dmax" mi"
2710 MOVE 4500;1500: PRINT "SS 1/4 mi: "et4"s/"v4"mi/h"
2720 MOVE 4500;1250: PRINT "0-60mi/h:  "t60"s"
2725 MOVE 900;400: PRINT n$": "d1$
2730 MOVE 5800;400: PRINT "Press c to continue."
2740 IF INKEY$<>"c" THEN 2740
2750 RETURN

2900 LABEL inpw
2910 CLS: WINDOW #1 TITLE "DIMENSIONS AND LAUNCH DATA"
2912 IF wt=0 THEN 2920
2914 PRINT AT (10) "Is the data (weight, wheelbase, etc) the same a
s before? y/n: ",
2914  REPEAT: inp$=INKEY$: UNTIL inp$>"": IF inp$="y" THEN 3007
2920 INPUT AT(10) "Weight of bike and rider, lbf: ",wt:m=wt/32.2: P
RINT
2930 PRINT TAB(10) "If the following data is not available,"
2940 PRINT TAB(10) "enter 0; the program will assume a typical valu
```

```
e."
2950 PRINT
2960 INPUT AT(10) "Wheelbase, inches: ",wb: IF wb=0 THEN wb=56
2970 PRINT TAB(10) "Centre of gravity co-ordinates"
2980 INPUT AT(10) "Horizontal distance from rear wheel spindle, inc
hes: ",cgx: IF cgx=0 THEN cgx=wb*0.5
2990 INPUT AT(10) "Height above ground, inches: ",cgy: IF cgy=0 THE
N cgy=2*r+3: PRINT
3000 INPUT AT(10) "Coefficient of friction, tyre/road: ",mu: IF mu=
0 THEN mu=1: PRINT
3002 PRINT AT(10)"(Note: to start in a gear other than first, alter
 line 3215.)": INPUT AT(10) "At what engine speed is the clutch en
gaged? ",cl: IF cl=0 THEN cl=n(0)
3003 IF cl>0 AND cl<n(0) THEN CLS: PRINT AT(10) "This is less than
the first speed value.": PRINT AT(10) "Enter 0 or a value greater t
han "n(0): GOTO 3002
3005 PRINT AT(10) "At what engine speed are gearshifts made? "
3006 INPUT AT(10) "If 0 is entered, the program will calculate opti
mum speeds. ",gs
3007 CLS: GOSUB spec
3010 RETURN


3100 LABEL avg
3105 FOR i=0 TO 220: warn(i)=0: thr(i)=0: FOR ii=1 TO y: fav(ii,i)=
0: NEXT: NEXT

3110 i=1
3120 REPEAT: j=0
3130  REPEAT
3135   FOR v=vr(i,j) TO vr(i,j+1)
3140   fav(i,v)=f(i,j)+(f(i,j+1)-f(i,j))*(v-vr(i,j))/(vr(i,j+1)-vr(
i,j))
3145
3150   NEXT
3160  j=j+1
3170  UNTIL j-1=w-2
3200 i=i+1:IF i-1=y THEN 3210
3205 IF vr(i-1,j-1)<vr(i,0)-1 THEN GOSUB slip
3210 UNTIL i-1=y:
3215 sg=1: REM gear in which start is made
3218 speed=CEILING(cl*0.00595*r/(p*g(sg)*t2/t1))
3220 FOR v=0 TO speed
3225 fav(sg,v)=fav(sg,speed): warn(v)=warn(v)+0.5
3226 IF fav(sg,v)>0.8*f(sg,12) THEN fav(sg,v)=0.8*f(sg,i2)
```

```
3227 NEXT
3228 IF gs<>0 THEN GOSUB shift: GOTO 3340: REM condition for pre-ch
osen gearshift rpm.
3230 i=sg: v=0
3240   REPEAT
3250   REPEAT
3252   thr(v)=fav(i,v)
3255   GOSUB wheel
3260   thr(v)=thr(v)-(a+b*v+c*v^2)
3270   v=v+1
3275   UNTIL fav(i,v-1)<=fav(i+1,v-1) OR v-1=vr(i,w-2)
3280   shift(i)=v-1: vsh(i)=shift(i)*p*g(i)*t2/(t1*0.00595*r)
3290   i=i+1
3295   UNTIL i-1=y OR thr(v-1)<=0
3300
3310
3340 RETURN


3400 LABEL shift
3410 i=sg:v=0
3420 FOR gear=i TO y-1: shift(gear)=ROUND(gs*0.00595*r/(p*g(gear)*t
2/t1)): vsh(gear)=gs: NEXT
3430 REPEAT
3440 REPEAT
3441 IF fav(i,v)=0 THEN fav(i,v)=f(i,0): IF FRAC( warn(v))<>0.5 THE
N warn(v)=warn(v)+0.5
3442 thr(v)=fav(i,v)
3445 GOSUB wheel
3450 thr(v)=thr(v)-(a+b*v+c*v^2): v=v+1
3460 UNTIL v-1=shift(i)
3470 i=i+1
3480 UNTIL i-1=y OR thr(v-1)<=0
3490 RETURN

3500 LABEL slip
3510 FOR v=vr(i-1,j-1) TO vr(i,0)-1
3520 fav(i,v)=f(i,0): warn(v)=warn(v)+0.5
3530 NEXT
3540 RETURN


3600 LABEL tyre
3610 WINDOW #2 FULL ON: CLS #2: WINDOW #2 TITLE "Tyre size and roll
```

```
ing radius": WINDOW #2 OPEN
3620 SET ZONE 19
3625 PRINT #2 " Rolling radius in inches ±2%, based on ETRTO standa
rd."
3630 PRINT #2: PRINT #2 " 15-inch","17-inch","18-inch","18-inch"
3640 PRINT #2
3650 PRINT #2 "140/90  11.96","2.50     10.73","3.60     11.62","100/
80  11.66"
3660 PRINT #2 "150/90  12.30","2.75     11.11","4.10     12.11","110/
80  11.96"
3670 PRINT #2 " ","3.00     11.38","4.25     12.91","120/80  12.26"
3680 PRINT #2 " 16-inch","4.50     12.59","4.25/85 12.46","130/80  1
2.57"
3690 PRINT #2 " "," "," "," "

3700 PRINT #2 "4.60     11.42","100/80  11.18","90/90     11.7","110/7
0  11.55"
3710 PRINT #2 "100/90  11.07","120/80  11.79","100/90  12.04","140/
70  12.34"
3720 PRINT #2 "110/90  11.42"," ","110/90  12.38","150/70  12.61"
3730 PRINT #2 "120/90  11.75","110/90  11.90","120/90  12.72",""
3740 PRINT #2 "130/90  12.09","120/90  12.24","130/90  13.06"," "
3750 PRINT #2 "140/90  12.44","130/90  12.59","140/90  13.40","170/
60  12.49"
3760 PRINT #2 "100/80  10.70"," "
3770 PRINT #2 "120/80  11.30","140/80  12.39"
3780 PRINT #2 "150/80  12.21"
3790 PRINT #2 TAB(40) "Press c to continue."
3800 IF INKEY$<>"c" THEN 3800
3810 WINDOW #1 OPEN
3820 RETURN

4000 LABEL pri
4050 GOSUB setup
4010 LPRINT n$, d1$: LPRINT
4020 LPRINT "Final drive "t2"/"t1: LPRINT
4030 LPRINT "speed, mi/h","thrust, lbf": LPRINT
4040 FOR j=1 TO y
4050 FOR i=0 TO w-2
4060 LPRINT ROUND(v(j,i),1),ROUND(f(j,i),1)
4070 NEXT: LPRINT "Gear "j": " ROUND(v(j,i-1)*1000/n(i-1),1)" mi/h
per 1000rev/min": LPRINT
4080 NEXT
4090 LPRINT: LPRINT
4100 RETURN
```

```
4850 LABEL pow
4860 pmax=hp(0):tmax=t(0):i=0
4870 REPEAT: i=i+1: pmax=MAX(pmax,hp(i)): UNTIL i=w-1: i=0
4875 REPEAT: i=i+1: tmax=MAX(tmax,t(i)): UNTIL i=w-1: mmax=MAX(pmax
,tmax)
4880 sx=6500/n(w-2)
4890 sy=3800/mmax
4900 CLS: WINDOW #1 TITLE "Output (bhp and torque, lb ft) v. crank
speed, rev/min x10^3"
4910 k=1000: i=0
4920 REPEAT: LINE 1000+k*i*sx;1000,1000+k*i*sx;5000: MOVE 800+k*i*s
x;600: PRINT i: i=i+1
4930 UNTIL k*(i-1)*sx >=6500
4940 i=0: IF sy<390 THEN k=10 ELSE k=1
4950 REPEAT: LINE 1000;1000+k*i*sy,8000;1000+k*i*sy: MOVE 400;1000+
k*i*sy: PRINT i*k: i=i+1: UNTIL k*i*sy>=3900
4960 FOR i=0 TO w-3: LINE 1000+n(i)*sx;1000+hp(i)*sy,1000+n(i+1)*sx
;1000+hp(i+1)*sy: NEXT
4970 FOR i=0 TO w-3: LINE 1000+n(i)*sx;1000+t(i)*sy,1000+n(i+1)*sx;
1000+t(i+1)*sy: NEXT
4980 MOVE 1000;300: PRINT n$": "d1$
4990 MOVE 5800;300: PRINT "Press c to continue."
5000 IF INKEY$ <>"c" THEN 5000
5010 RETURN

6000 LABEL quart
6010 et4=et: v4=i: d4=dist: counter=0: ii=i
6020 REPEAT
6030 v4=v4 + 0.1: counter=counter + 1
6040 et4=et4 + 0.1/a(ii)
6050 d4=d4+ v4/(a(ii)*36000)
6060 IF counter=9 THEN counter=-1 AND ii=ii+1
6070 UNTIL d4>0.2498
6080 RETURN

6100 LABEL nam
6110 n1$="": n2$=".RL1"
6120 le=LEN(n$)
6130 FOR i=1 TO le
6140 t$=n$(i)
6150 IF ASC(t$)=32 THEN t$=""
6160 n1$=n1$+t$
6170 NEXT
6180 le=LEN(n1$)
6190 IF le>8 THEN n1$=n1${TO 3}+n1${-5 TO}
```

230

```
6195 n1$=n1$+n2$
6200 RETURN

6300 LABEL fil
6310 CLS: WINDOW #1 TITLE "Saving data on disc file"
6320 PRINT AT(10;3)"Put a formatted disc in drive B:"
6330 PRINT: PRINT AT(10)"The data will be stored in \RL\"n1$
6340 PRINT: PRINT AT(10)"To alter the filename, press A or press"
6342 PRINT AT(10) "any other key to continue."
6350 REPEAT:f1$=INKEY$: UNTIL f1$>"": IF LOWER$(f1$)="a" THEN GOSUB
  ns
6360 DRIVE "B"
6370 ON ERROR GOTO 6550
6380 CD \RL\
6390
6400 ON ERROR GOTO 0
6410
6420 IF n1$="" THEN n1$=n3$
6430 IF FIND$(n1$)>"" THEN REPEAT: GOSUB rname: UNTIL FIND$(n1$)=""
  OR r$="y"
6450 OPEN #3 OUTPUT n1$
6460 PRINT #3,n$
6462 PRINT #3,d1$
6470 PRINT #3,w,y,p,t1,t2,r,wt,wb,cgx,cgy,mu,cl,gs
6480 FOR i=1 TO y
6490 PRINT #3,g(i)
6500 NEXT
6510 FOR i=0 TO w-2
6520 PRINT #3,n(i),hp(i),t(i)
6530 NEXT
6540 CLOSE #3: GOTO 6570
6550 IF ERR=133 THEN GOSUB drctry
6560 RESUME NEXT
6570 RETURN

6600 LABEL rname
6610 PRINT
6620 PRINT AT(10)"A file "n1$" already exists."
6630 INPUT AT(10)"Do you want this file to replace it? y/n ",r$
6640 IF r$="y" THEN 6670
6650 mm=ASC(n1$(-1))+1
6660 n1$(-1)=CHR$(mm)
6670 RETURN

6700 LABEL drctry
6710 MD \rl\
```

```
6720 CD \rl\
6730 RETURN


6800 LABEL rea
6810 CLS: WINDOW #1 TITLE "READ DATA FROM FILE"
6820 DRIVE "B"
6830 PRINT AT(10;6)"Put the file disc into drive B:"
6840 PRINT AT(10)"Press a key when ready."
6850 IF INKEY$="" THEN 6850
6855 ON ERROR GOTO 6990
6860 CD \rl\
6870 ON ERROR GOTO 0
6880 PRINT: PRINT AT(10)"Note that file names are compressed into a
 maximum": PRINT AT(10)"of eight characters, with no spaces, and ha
ve ": PRINT AT(10)"an extension .RLx":GOSUB look
6890 PRINT: REPEAT: INPUT "Filename for machine, or Q to quit: ",n3
$
6895 IF n3$="q" THEN 6905
6900 IF FIND$(n3$)="" THEN PRINT "No file for "n3$
6905 UNTIL FIND$(n3$)>"" OR n3$="q": IF n3$="q" THEN 7000
6910 PRINT: PRINT "Reading "n3$
6920 OPEN #3 INPUT n3$
6930 INPUT #3,n$
6932 INPUT #3,d1$
6940 INPUT #3,w,y,p,t1,t2,r,wt,wb,cgx,cgy,mu,cl,gs
6950 FOR i=1 TO y
6960 INPUT #3,g(i)
6965 NEXT
6970 FOR i=0 TO w-2
6975 INPUT #3,n(i),hp(i),t(i)
6980 NEXT
6985 CLOSE #3
6986 m=wt/32.2
6987 CLS:PRINT: PRINT AT(10)n$
6988 tim=TIME:REPEAT:UNTIL TIME>tim+500:GOTO 7020


6990 IF ERR=133 THEN PRINT AT(10)"There is no \RL\ directory on thi
s disc."
7000 PRINT TAB(10)"Try another disc or quit? a/q ";: INPUT re$
7010 IF re$="a" THEN 6810
7020 RETURN


7100 LABEL ns
7110 PRINT "Existing filename: "n1$
7120 INPUT "New filename, including extension: ",n1$
```

232

```
7130 PRINT "Existing full name: "n$
7140 INPUT "New name: ",n$
7150 RETURN


7200 LABEL wheel
7210 IF thr(v)*cgy>wt*cgx THEN warn(v)= warn(v)+1: thr(v)=wt*cgx/cg
y
7220 IF thr(v)>wt*mu THEN warn(v)=warn(v)+2: thr(v)=wt*mu
7230 RETURN


7400 LABEL warn
7405 WINDOW #1 TITLE "Gearshift and traction data"
7410 CLS
7415 sxx=6300/vmax
7420 FOR i=0 TO vmax
7430 IF INT(warn(i))=1 OR INT(warn(i))=3 THEN LINE(1500+i*sxx);1000
,(1500+(i+1)*sxx);1000 WIDTH 5 COLOUR 1
7440 IF INT(warn(i))=2 OR INT(warn(i))=3 THEN LINE(1500+i*sxx);1500
,(1500+(i+1)*sxx);1500 WIDTH 5 COLOUR 2
7450 IF FRAC(warn(i))=0.5 THEN LINE(1500+i*sxx);2000,(1500+(i+1)*sx
x);2000 WIDTH 5 COLOUR 11
7460 NEXT
7470 LOCATE 1;17: PRINT "Wheelie"
7480 LOCATE 1;15: PRINT "Wheelspin"
7490 LOCATE 1;13: PRINT "Clutch slip"
7500 FOR i=0 TO vmax STEP 10
7510 LINE(1500+i*sxx);1000,(1500+i*sxx);2100
7520 MOVE (1300+i*sxx);700: IF i<vmax-15 THEN  PRINT i
7530 NEXT
7540 LOCATE 10;1: PRINT "Gearshift   ","rev/min","mi/h"
7545 IF gs=0 THEN PRINT AT(10) "(calculated)"
7550 PRINT
7560 FOR i=1 TO y-1
7570 PRINT AT(10) i" to "i+1,ROUND(vsh(i)),shift(i)
7580 NEXT
7580 PRINT AT(12;19)"Road speed, mi/h" AT(50;19) "Press any key to
continue"
7590 IF INKEY$="" THEN 7590
 7600 RETURN


8000 LABEL look
8010 PRINT "Here are the files on this disc:"
8020 FILES
8030 RETURN
```

```
8100 LABEL setup
8110 LPRINT CHR$(15)
8120 LPRINT CHR$(27)+"1"+CHR$(15)
8130 LPRINT CHR$(27)+"N"+CHR$(5)
8140 RETURN

8200 LABEL spec
8202 WINDOW #2 SIZE 27,30: WINDOW #2 PLACE 400,0
8203 WINDOW #2 TITLE "CURRENT SPECIFICATION"
8210 CLS #2: PRINT #2
8220 PRINT #2 " Final drive      "t2"/"t1
8230 PRINT #2 " Primary          "ROUND(p,3)
8240 PRINT #2 " Roll radius      "r
8250 PRINT #2
8260 PRINT #2 " Wheelbase        "wb
8270 PRINT #2 " cgx - r wheel    "cgx
8280 PRINT #2 " cgy              "cgy
8290 PRINT #2 " Total weight     "wt
8300 PRINT #2 " µ                "mu
8310 PRINT #2
8320 PRINT #2 " Start gear       "sg
8330 PRINT #2 "Clutch dump rpm "cl
8340 PRINT #2
8350 PRINT #2 " Total drag"
8360 PRINT #2 "   =a + bv + cv² "
8370 PRINT #2 "        a="a
8380 PRINT #2 "        b="b
8390 PRINT #2 "        c="c
8400 RETURN
```

This program, written in BASIC2, was developed simply as a device to take engine power curves and work out the corresponding thrust that this would produce at the back wheel, in the various gears. Since then it has had bits added to it as needed and now it predicts top speed, compares aerodynamic features, takes weight and traction data to work out acceleration, finds optimum gear-shift revs, warns if wheelspin or wheelies are likely, can store all the data in a disc file and read it back again.

It needs power figures and engine speeds, and details of the gearing and tyre size. This it converts into cascade curves showing the thrust at the back wheel in each gear, against road speed. It adds a graph of total resistance (air drag, rolling drag, driveline drag and inertia); as long as the bike has more thrust than this, it has power to accelerate to a higher speed. Where the two lines cross is the maximum speed the bike can reach or, if it is under-geared, it will reach the end of its rev range in top gear before meeting the drag line.

So the first function is to see the effects of different gearing (and tyre size). The figures for total drag are empirical (see program AERO) and are simply devised to fit the measured performance of a wide selection of bikes. Total drag is assumed to take the form:

$$\text{Drag} = a + bv + cv^2 + ...$$

Where a, b, c are constant (for that bike) and v is its road speed. Any values for a, b, c can be fed into the program. Values which tally with the performance of a similar bike can be used and, if the subject bike then exceeds this performance, it obviously has less drag, and vice versa.

Now, if the drag force is subtracted from the available thrust, what is left is the force which accelerates the bike. So if the computer is told the weight of the bike it can work out the acceleration. If it is also told the weight distribution and the tyre friction, it can tell whether there is enough power to lift the front wheel or spin the back one.

What it does is to go from zero speed in 1 mph steps, work out the available thrust and the acceleration at each speed and, from this it calculates the time taken to reach that speed and the distance covered. The points, taken at probably 500 rpm intervals, would only appear at similar intervals on the graph. In top gear there might be gaps of 7 mph between them. The computer works out a straightline average from one point to the next, in order to get a thrust value at 1 mph intervals.

Of course there are still gaps. The first thrust value does not appear until the bike is doing perhaps 30 mph, so the computer assumes that you slip the clutch until you reach this speed, takes a percentage of the first thrust value (or whatever engine speed you tell it you pull away at) and uses this value. If that exceeds the weight transfer on the back wheel multiplied by the tyre's coefficient of friction it treats that as a limiting value and posts a 'wheelspin' warning. The tyre friction is found empirically, like drag, and 0.8 works quite well for road tyres, 1.0 or more for race tyres on a dry surface.

If the weight transfer is enough to turn the bike over backwards, the program holds this as a limiting value and posts a 'wheelie' warning. Unlike most riders, the computer can easily hold the engine on the point of wheelspin, with the front wheel a steady inch above the floor and change up a gear when necessary.

It shifts into the next gear either at the engine speed you tell it to, or it works out the optimum speed for itself. If it reaches the last thrust value in the first gear or if the thrust value in the second gear is higher than the value for the first, it changes up. If there is a gap between the gears, it reaches a road speed for which there is no thrust value (i.e., the power band is too narrow or the gear spacing too wide) so, on coming to the end of the first gear it shifts up and then slips the clutch, holding a steady engine speed and taking a percentage of the first thrust value in the next gear (to approximate for the power lost in slipping the clutch). From the acceleration figures, it draws a graph of road speed against time, plucking out 0–60 mph and quarter mile figures. It will display optimum gearshift points, and indicate at which road speeds there is enough power to spin the back wheel or lift the front. Finally, it indicates over which speed range(s) it is necessary to slip the clutch.

The effects on top speed and acceleration can quickly be seen if you change the gearing, tyre size, tyre grip, weight, centre of gravity and aerodynamics. By playing about with these variables you can discover what has to be done to increase performance and what would be a waste of time. As the bike is tested and more data is gathered, the predictions become more accurate.

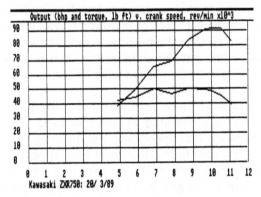

Figure A.5 The program RL.BAS takes power figures as its input and produces bhp and torque curves ...

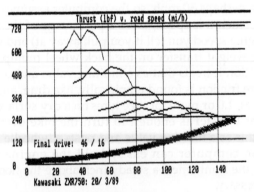

Figure A.6 ... which it translates into curves representing thrust at the rear wheel in each gear. The broad line equates to the total drag acting on the bike

236

On an untried 250 the program was within 1.5 mph (1.2 per cent) of the top speed but a long way out on quarter mile times, where the error was 7.4 per cent, partly because the riders found the peaky two-stroke harder to get off the line than the computer did. On the GSX-R1100 Suzuki and its race-modified variants there was a lot of test data and the predictions were close: a tenth of a second on the quarter mile times (1 per cent error) and 2 mph in a top speed of 184 mph (1.1 per cent error).

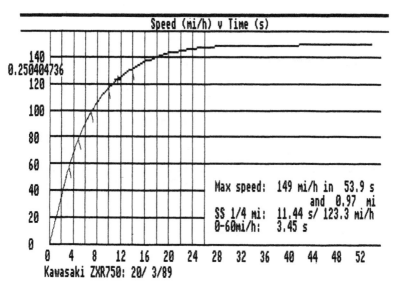

Figure A.7 The difference between thrust and drag (in Figure A.6) is the force which can accelerate the bike, and the program produces this acceleration curve. The measured performance of the ZXR750 was 152.5 mph maximum speed and 11.4 s/120.5 mph for the standing-start quarter mile. *(Performance Bikes, June 1989)*

There can be a minor glitch in the program when it scales the x- and y-axes before drawing a graph. It selects the maximum values in the data and then scales the grid to use the full dimensions of the screen. Sometimes it tries to print outside the screen and the computer puts up an error message like 'co-ordinate outside virtual screen', returning to the edit mode at the line where this happened. Look at window #1. If vertical and horizontal lines have been drawn, the fault is in the y-axis scale. If only vertical lines have appeared, the fault is in the x-axis scale. Go back to the edit mode, and a few lines before the fault there will be a line which defines sx and sy, the x- and y-scale factors. If the x-axis is at fault, then change the line from sx = 7000 to sx = 6000; to change the scale of the y-axis, alter the value for sy. To restart the program without losing any of the data, return to the command mode/dialogue box and type GOTO 148 [enter]. This will take the program to the main menu. You will have to make window #1 visible in order to see the menu.

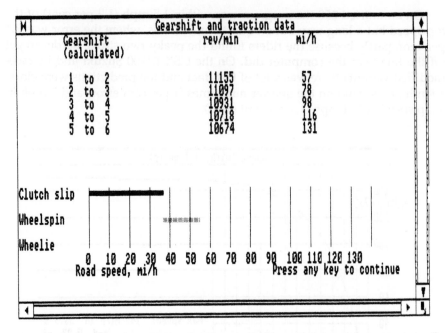

Figure A.8 The program also calculates optimum gearshift speeds, where it is necessary to slip the clutch and the limiting condition where the machine has enough power to spin the rear wheel (38 to 55 mph in this case) or lift the front wheel.

Program AERO

Analysis of terminal speed versus power required.

```
10 REM aero
20 REM terminal speed v power required
30 REM 3/6/88: 7/7/89
40 REM jwr
50 REM

100 x=1000: y=3: item =0: ma=1: maa=2: si=1: co=1
105 nam$="AERO.DAT"
110 OPTION DATE 1: OPTION DEGREES
120 DIM n$(x), d(x)
130 DIM v(x),p(x),w(x),wang(x), peak(x),f(x)

200 GOSUB show
202 GOSUB head
205 GOSUB inp
210 REPEAT
220 GOSUB menu
230 UNTIL aa>=9
```

238

```
235 USER ORIGIN 0;0
240 END


300 LABEL show
310 WINDOW #1 FULL
320 WINDOW #2 SIZE 25,15
330 WINDOW #2 PLACE 400,0
340 RETURN


400 LABEL menu
402 WINDOW #1 OPEN
405 CLS #2: WINDOW #2 OPEN: WINDOW #2 TITLE "Menu"
406 PRINT #2
410 PRINT #2 "  Enter new data....1"
420 PRINT #2 "  Display data/bhp..2"
425 PRINT #2 "         /lb thrust..3"
430 PRINT #2 "  Draw graphs.......4"
431 PRINT #2 "  File data.........5"
433 PRINT #2 "  Review data.......6"
434 PRINT #2 "  Alter graph line..7"
435 PRINT #2 "  Select by year....8"
436 PRINT #2 "  STOP..............9"
440 PRINT #2
450 REPEAT: aa=INKEY: UNTIL aa>-1: aa=aa-48
455 IF aa>9 OR aa<1 THEN 450
460 ON aa GOSUB inp,disp,dispf,grap,fil,rev,lin,sel
470 RETURN


500 LABEL fil
510 CLS: WINDOW #1 OPEN: WINDOW #1 TITLE "Power v. speed"
515 GOSUB dis
600 IF v(1)=0 THEN GOSUB inp
610 OPEN #y OUTPUT nam$
620 FOR i=1 TO item
630 PRINT #y,d(i),v(i),p(i),w(i),wang(i),n$(i)
660 NEXT
670 CLOSE #y


770 IF ERR=133 THEN GOSUB drctry: RESUME NEXT
780 RETURN
781 LABEL dis
782 PRINT AT(10) "Put the file disc, or a formatted disc, into
    drive B:"
783 PRINT AT(10) "Press any key when ready."
784 REPEAT: a$=INKEY$: UNTIL a$>""
```

```
785 PRINT
786 DRIVE "B"
787 er=0
788 ON ERROR GOTO 770
789 CD \AERO
790 ON ERROR GOTO 0
791 RETURN


800 LABEL drctry
810 MD \AERO
820 CD \AERO
830 RETURN


900 LABEL inp
910 CLS: WINDOW #1 OPEN: WINDOW #1 TITLE "Speed/power input."
920 PRINT TAB(10)"Use mi/h and bhp."
930 PRINT TAB(10)"Wind angle: headwind = 0 deg, tailwind = 180
 deg"
940 PRINT: PRINT
942 PRINT TAB(10) "Do you want to create a new file"
944 PRINT TAB(10) "or read an existing file? n/e "
946 REPEAT: a$=INKEY$: UNTIL a$>"": IF a$="n" THEN GOSUB cre E
LSE GOSUB rea
960 i=0
950 REPEAT: i=i+1: UNTIL v(i)=0
955 PRINT TAB(10)i-1" entries read from disc file. "
956 PRINT TAB(10)"Do you want to add any more? y/n"
957 REPEAT: a$=INKEY$: UNTIL a$>"": IF a$="n" THEN CLS: PRINT
AT(10) "Now use MENU >>>>": GOTO 959
958 GOSUB wri
959 RETURN


960 LABEL wri
962 CLS
965 REPEAT
970 INPUT "Model: ",n$(i)
980 INPUT "Model year (last two digits only): ",d(i): IF d(i)>
1000 THEN d(i)=d(i)-1900
990 INPUT "Maximum speed: ",v(i)
1000 INPUT "Power: ",p(i)
1010 INPUT "Wind speed: ",w(i)
1020 INPUT "Wind direction: ",wang(i): PRINT
1030 i=i+1
1040 PRINT "Another entry? y/n "
1050 REPEAT: a$=INKEY$: UNTIL a$>""
```

```
1060 UNTIL a$="n"
1065 item=i-1
1066 PRINT TAB(10) "To file the new data, use menu option 5"
1067 PRINT: PRINT TAB(10)"Press any key to continue."
1068 REPEAT: a$=INKEY$: UNTIL a$>""
1070 RETURN


1100 LABEL rea
REM drive b:/ insert disc/ cd \aero\ /read file into array
1110 CLS #2: WINDOW #2 OPEN: WINDOW #2 TITLE "Read disc file"
1120 PRINT #2
1130 PRINT #2 " Put the file disc "
1140 PRINT #2 " into drive B:": PRINT
1150 PRINT #2 " Press any key when ready"
1160 REPEAT: a$=INKEY$: UNTIL a$>""
1170 DRIVE "B"
1175 ON ERROR GOTO 1300
1180 CD \AERO
1190 ON ERROR GOTO 0
1192 PRINT #2 "File name?"
1193 PRINT #2"...defaults to AERO.DAT"
1194 PRINT #2 "if you press ENTER": INPUT #2 nam$: IF nam$=""
THEN nam$="AERO.DAT"
1200 PRINT #2 " Reading "nam$
1210 OPEN #y INPUT nam$
1220 i=item
1230 WHILE NOT(EOF(#y))
1240 i=i+1
1250 INPUT #y,d(i),v(i),p(i),w(i),wang(i),n$(i)
1260 WEND
1270 CLOSE #y
1275 item=i
1280 GOTO 1330

1300 IF ERR=133 THEN PRINT #2 "No AERO directory": PRINT #2 "o
n this disc.": PRINT #2 "Try another disc ": PRINT #2 "or quit
? a/q"
1310 REPEAT: a$=INKEY$: UNTIL a$>"": IF a$="a" THEN CLS #2: GO
TO 1120
1320 CLOSE
1330 RETURN


1400 LABEL disp
1410 CLS #1: WINDOW #1 OPEN
```

```
1420 WINDOW #1 TITLE "AERO.BAS"
1425 unit$="p"
1430 IF v(1)=0 THEN GOSUB rea
1435 IF aa=2 THEN GOSUB clean
1440 i=0:vmax=0: pmax=0: REPEAT: i=i+1
1450 IF v(i)>vmax THEN vmax=v(i)
1460 IF p(i)>pmax THEN pmax=p(i)
1470 UNTIL v(i)=0
1480 sx=7000/(1.1*pmax)
1490 sy=4000/(1.1*vmax)
1500 USER ORIGIN 1000;1000
1505 IF vmax<120 THEN k=10 ELSE k=20
1510 FOR j=0 TO vmax STEP k
1520 LINE 0;j*sy,7000;j*sy
1530 MOVE -500;j*sy-70: PRINT j
1540 NEXT
1545 IF pmax<100 THEN k=10 ELSE k=20
1550 FOR j=0 TO pmax STEP k
1560 LINE j*sx;0,j*sx;4000
1570 MOVE j*sx-250;-400: PRINT j
1580 NEXT
1590 MOVE -1000;3500: PRINT "mi/h": IF aa>3 THEN 1600
1595 FOR j=1 TO 4: PLOT 50;4000-j*300 MARKER j+1:MOVE 150;4000
-j*300: PRINT 75+j*3"-"77+j*3" model": NEXT
1600 MOVE 0;-700: PRINT "bhp"
1610 IF aa>3 THEN GOSUB yr
1620 FOR j=1 TO i-1
1625 IF aa>3 THEN m=maa: GOTO 1665
1630 IF d(j)<=80 THEN m=2
1640 IF d(j)>80 AND d(j)<84 THEN m=3
1650 IF d(j)>83 AND d(j)<87 THEN m=4
1660 IF d(j)>86 THEN m=5
1665 IF peak(j)=1 THEN 1680
1670 PLOT p(j)*sx;v(j)*sy MARKER m
1680 NEXT
1685
1690
1695 RETURN

1700 LABEL grap
1701 CLS #2: WINDOW #2 OPEN
1703 WINDOW #2 TITLE "Drag factors"
1705 PRINT #2 "Do you want to use"
1706 PRINT #2 "bhp or lbf thrust? b/t"
1710 REPEAT: a$=INKEY$: UNTIL a$>"": IF a$="b" THEN GOSUB grap
a: GOTO 1900
```

```
1715 GOSUB dispf: WINDOW #2 OPEN: CLS #2
1720 PRINT #2
1730 PRINT #2 " Overall drag is ": REM    this is calculated in
1740 PRINT #2 " assumed to take": REM     lbf thrust, not bhp
1750 PRINT #2 " the form:"
1760 PRINT #2 " drag = a + bv + cv²"
1770 PRINT #2
1780 PRINT #2 "Options used in RL:"
1790 PRINT #2 "1. a=16, b=0, c=0.0105"
1800 PRINT #2 "2. a=16, b=0, c=0.0091"
1810 PRINT #2 "3. a=16, b=0, c=0.0080"
1820 PRINT #2 "4. new values"
1830 PRINT #2 "    Option? ..."
1840 REPEAT: z=INKEY: UNTIL z>-1: z=z-48
1850 IF z=1 THEN a=16 : b=0 : c=0.0105
1860 IF z=2 THEN a=16 : b=0 : c=0.0091
1870 IF z=3 THEN a=16 : b=0 : c=0.0080
1880 IF z=4 THEN INPUT #2 "a = ",a: INPUT #2 "b = ",b: INPUT #
2 "c = ",c
1890 GOSUB draw
1895 PRINT #2 "Another line? y/n"
1896 REPEAT: a$=INKEY$: UNTIL a$>""
1897 IF a$="y" THEN PRINT #2 " Option? ...": GOTO 1840
1900 RETURN

2000 LABEL dispf
2050 CLS #1: WINDOW #1 OPEN: WINDOW #1 TITLE "AERO.BAS"
2010 IF v(1)=0 THEN GOSUB rea
2015 IF aa=3 THEN GOSUB clean
2020 fmax=0: unit$="f"
2030 FOR i=1 TO item
2040 f(i)=p(i)*375/v(i)
2050 IF f(i)>fmax THEN fmax=f(i): IF v(i)>vmax THEN vmax=v(i)
2060 NEXT
2070 sfx=7000/(fmax)
2080 sy=4000/(1.1*vmax)
2090 USER ORIGIN 1000;1000
3000 IF vmax<120 THEN k=10 ELSE k=20
3010 FOR j=0 TO vmax STEP k
3020 LINE 0;j*sy,6700;j*sy
3030 MOVE -400;j*sy-70: PRINT POINTS(8) j
3040 NEXT
3045 IF fmax<100 THEN k=10 ELSE k=20
3050 FOR j=0 TO fmax STEP k
3060 LINE j*sfx;0,j*sfx;3700
```

```
3070 MOVE j*sfx-200;-350: PRINT POINTS(8) j
3080 NEXT
3090 MOVE -1000;3500: PRINT "mi/h": IF aa>3 THEN 3100
3095 FOR j=1 TO 4: PLOT 50;4000-j*300 MARKER j+1:MOVE 150;4000
-j*300: PRINT 75+j*3"-"77+j*3" model": NEXT
3100 MOVE 0;-700: PRINT "lbf"
3110 IF aa>3 THEN GOSUB yr
3120 FOR j=1 TO item
3125 IF aa>3 THEN m=maa: GOTO 3165
3130 IF d(j)<=80 THEN m=2
3140 IF d(j)>80 AND d(j)<84 THEN m=3
3150 IF d(j)>83 AND d(j)<87 THEN m=4
3160 IF d(j)>86 THEN m=5
3165 IF peak(j)=1 THEN 3180
3170 PLOT f(j)*sfx;v(j)*sy MARKER m
3180 NEXT
3185
3190
3200 RETURN


3300 LABEL grapa
3320 PRINT #2: GOSUB disp: WINDOW #2 OPEN: unit$="p": CLS #2
3330 PRINT #2 " Overall drag hp is "
3340 PRINT #2 " assumed to take"
3350 PRINT #2 " the form:"
3360 PRINT #2 " drag = av + bv² + cvv²"
3370 PRINT #2
3380 PRINT #2 "Options used in RL:"
3390 PRINT #2 "a = 0.0427, b=0 and"
3300 PRINT #2 "opt 1: c = 279 x 10^-7"
3310 PRINT #2 "opt 2: c = 242 x 10^-7"
3315 PRINT #2 "opt 3: c = 213 x 10^-7"
3320 PRINT #2 "opt 4: new values"
3330 REPEAT: PRINT #2 "    Option? ..."
3340 REPEAT: z=INKEY: UNTIL z>-1: z=z-48
3350 IF z=1 THEN a=0.0427: b=0 : c=0.0000279
3360 IF z=2 THEN a=0.0427: b=0 : c=0.0000242
3370 IF z=3 THEN a=0.0427: b=0 : c=0.0000213
3380 IF z=4 THEN INPUT #2 "a = ",a: INPUT #2 "b = ",b: INPUT #
2 "c = ",c
3385 GOSUB draw
3386 PRINT #2 "Another line? y/n":REPEAT: a$=INKEY$: UNTIL a$>
""
3387 UNTIL a$="n"
3390 RETURN
```

244

```
3400 LABEL draw
3410 IF unit$="f" THEN xmax=fmax ELSE xmax=pmax
3420 IF unit$="f" THEN ssx=sfx ELSE ssx=sx
3430 i=0: REPEAT
3440 IF unit$="f" THEN p=a+b*i+c*i^2 ELSE p=a*i+b*i^2+c*i^3
3450 PLOT p*ssx;i*sy MARKER ma SIZE si COLOUR co
3460 i=i+1
3470 UNTIL p>xmax OR i>vmax
3475 MOVE 4000;1200: IF unit$="f" THEN PRINT"drag = a + bv + c
v²" ELSE PRINT "drag hp = av + bv^2 + cv^3"
3476 MOVE 4000;900: PRINT "where:  a = "a
3477 MOVE 4000;600: PRINT "        b = "b
3478 MOVE 4000;300: PRINT "        c = "c
3480 RETURN


3500 LABEL rev
3510 CLS #1: WINDOW #1 OPEN: WINDOW #1 TITLE "AERO.BAS"
3520 IF v(1)=0 THEN GOSUB rea
3522 SET ZONE 9
3525 PRINT " ","model","year","max mi/h", "max bhp", "wind v/θ
"


3530 FOR i=0 TO item
3540 PRINT i,n$(i),d(i),v(i),p(i),w(i)"/"wang(i)
3550 GOSUB wait
3560 NEXT
3570 PRINT: PRINT "Do you wish to alter any of these? y/n"
3580 REPEAT: a$=INKEY$: UNTIL a$>""
3590 IF a$="y" THEN GOSUB alt
3600 RETURN


3700 LABEL alt
3705 REPEAT
3710 PRINT : INPUT "Enter the item number: ",q
3720 PRINT "Current model is "n$(q),: INPUT "New model: ",n$(q
)
3730 PRINT "Current year is ",d(q),: INPUT "New year: ",d(q)
3740 PRINT "Current speed is ",v(q),: INPUT "New speed: ",v(q)
3750 PRINT "Current power is ",p(q),: INPUT "New power: ",p(q)
3760 PRINT "Current wind speed is ",w(q),: INPUT "New wind spe
ed: ",w(q)
3770 PRINT "Current wind direction is ",wang(q),: INPUT "New d
irection (0=head): ",wang(q)
3780 PRINT: PRINT TAB(10)"Another? y/n"
3790 REPEAT: a$=INKEY$: UNTIL a$>""
```

```
3800 UNTIL a$="n"
3810 RETURN

3900 LABEL wait
3910 IF ww>0 THEN ww=ww+1: GOTO 3970
3920 yy=VPOS
3930 IF yy>750 THEN 3980
3940 PRINT AT(10;20) "Press SPACE bar to continue.";
3950 IF INKEY$<>" " THEN 3950
3960 ww=1: PRINT AT(10;20)"                          "
3970 IF ww=13 THEN ww=0: PRINT: GOTO 3940
3980 RETURN

4000 LABEL lin
4010 CLS #1: WINDOW #1 OPEN
4020 PRINT AT(10;4)"To alter the thickness and colour "
4030 PRINT AT(10)"of the drag line on the graph:"
4040 PRINT
4050 INPUT AT(10)"Default size is 1. New size: ",si
4060 INPUT AT(10)"Default colour is 1. New colour: ",co
4070 INPUT AT(10)"Default marker is 1. New marker: ",ma
4075 PRINT: PRINT AT(10)"Markers for individual models:"
4076 INPUT AT(10)"Default marker is 2. New marker: ",maa
4080 RETURN

4100 LABEL head
4110 CLS
4120 WINDOW #1 OPEN
4130 PRINT AT(30;5) COLOUR(10) POINTS(20) MODE(4) "   aero
"
4140 PRINT AT(30) COLOUR(10) POINTS(16) ADJUST(16) MODE(1) " p
ower v. drag "
4150 PRINT AT(30) COLOUR(10) POINTS(10) MODE(4) "

4160 PRINT AT(30) COLOUR(10) POINTS(8) MODE(4) "     J 1988   J
ohn Robinson   "
4170 PRINT AT(35;16) COLOUR(10) POINTS(10) MODE(3) "Press a ke
y."
4180 i=-1 : j=1
4190 REPEAT: j=j+i
4200 ELLIPSE 4340;2500,2000,1.2 WIDTH 5 COLOUR 1+j
4210 UNTIL INKEY>-1
4220 RETURN

4300 LABEL sel
4310 CLS
```

```
4320 IF v(1)=0 THEN GOSUB rea
4330 GOSUB clean
4340 INPUT "Time period: from model year (last two digits only
): ",yr1
4350 INPUT "... to model year: ",yr2
4360 FOR j=0 TO item
4370 IF d(j)<yr1 THEN peak(j)=1
4375 IF d(j)>yr2 THEN peak(j)=1
4380 NEXT
4390 PRINT: PRINT "Do you want to use power (bhp) or thrust (l
bf)? p/t"
4400 REPEAT: a$=INKEY$: UNTIL a$>""
4410 CLS
4420 IF a$="p" THEN GOSUB disp ELSE GOSUB dispf
4430
4440 RETURN

4500 LABEL clean
4510 FOR j=0 TO item: peak(j)=0: NEXT
4520 RETURN

4600 LABEL yr
4610 MOVE 150;3700
4620 IF yr1<yr2 THEN PRINT "Model year "yr1" to "yr2
4630 IF yr1=yr2 THEN PRINT "Model year "yr1"          "
4640 RETURN

4700 LABEL cre
4710 GOSUB dis
4720 INPUT "What is the file name? ", nam$
4730 i=0
4740 GOSUB wri
4750 RETURN
```

This BASIC2 program simply stores data in a disk file. It collects
maximum speed and the power measured when the engine is running at
that speed. It can also add figures for wind speed and direction. The
information is stored with the bike model name and model year.

It will then plot all of these points on a graph, working either in bhp versus road speed (mph) or in terms of rear wheel thrust (lbf) against road speed. It can select machines from certain years, or print points with different markers for different years.

Because BASIC2 is fairly compatible with MS.DOS, the files can be edited, SORTed and rearranged using MS.DOS commands. For example all the Suzuki GSX, GSX-R750 and 1100 models can be plucked out of the main file and copied into a file called GSX.DAT by using:

FIND "gsx*" aero.dat >gsx.dat

This can then be displayed as a graph showing all Suzuki's sports four-stroke development. There is a glitch because the FIND program writes a blank line, followed by the title line which you gave it and the BASIC program then tries to read these two gratuitous lines as data. So the first two lines have to be edited out; some text editors work all right, such as EDLIN, but others, like RPED, leave an end-of-file marker and then the BASIC program puts up an error message such as EOF MET and cannot read the rest of the file.

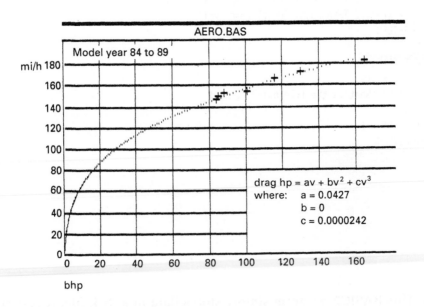

Figure A.9 (a)

248

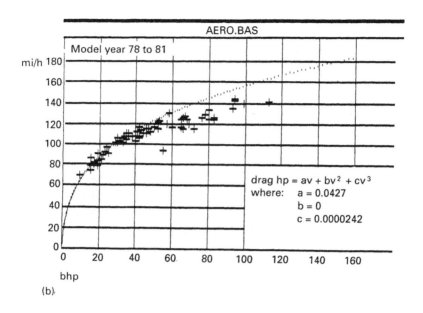

AERO.BAS

Model year 78 to 81

$$\text{drag hp} = av + bv^2 + cv^3$$
where: a = 0.0427
        b = 0
        c = 0.0000242

(b)

Figure A.9 (b)

Figure A.9 (c)

Figure A.9 (a) Using the program AERO.BAS, the maximum speed versus engine power can be plotted for various selections of machines. This represents seven variants of the Suzuki GSX-R models. The graph-drawing facility allows curves to be matched to the measured performance. This one fits the Suzukis fairly well. **(b)** The curve derived for the Suzuki GSX-R can be superimposed on a selection of other models to compare aerodynamic/total drag features. This shows a selection of bikes from 1978 to 1981 and clearly demonstrates the improvements made between that period and the development of the GSX-R **(c)** Suzuki's GSX-R models with standard and tuned engines provide a variety of power outputs with a fairly constant level of drag.

249

The more points there are, the more clearly the trend shows and it is possible to compare a given machine with its contemporaries or with machines from other years. The program also offers a graph-drawing facility, which will superimpose curves on the points already plotted.

These curves represent total drag (air drag, rolling drag, driveline drag and inertia in the moving parts) which is assumed to take the form:

Drag force = $a + bv + cv^2$

Or drag $hp = av + bv^2 + cv^3$

where any values for a, b and c can be inserted. Thus it is possible to find a curve which closely corresponds to the measured performance and to use this curve for comparative purposes here, or to use it in another program, such as RL above.

Program: brake force/spreadsheet

This program is based on a spreadsheet called AS-EASY-AS, designed for IBM PC and compatible computers running MS.DOS. It is made by Trius Inc. and distributed in the UK by Shareware Marketing.

The spreadsheet is based on a grid with its vertical columns labelled alphabetically (a, b, c, etc.) and its horizontal rows labelled numerically (1, 2, 3, etc.) from the top left corner, so the square which is four along and three down is d3.

Each box can contain:

- An input value which you type in.
- a label (words which appear on the screen) which is identified by ' at the beginning.
- A value, which is a number or begins with + or −, or a formula, which begins with a @.

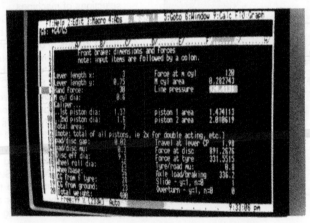

Figure A.10 The spreadsheet program which calculates braking forces

250

If box g6 contains the expression g4/g5 then this box has the result of dividing the contents of box g4 by box g5.

This spreadsheet contains the grid shown in Table A.1.

Table A.1 Spreadsheet calculations for brake forces

	Column a	c	e	g
1			'TITLE SPACE	
2				
3				
4	'Lever length x	INPUT VALUE	'Force at m cyl	c6*c4/c5
5	'Lever length y	INPUT VALUE	'M cyl area	@PI*c7^2/4 *see note 1*
6	'Hand force	INPUT VALUE	'Line pressure	g4/g5
7	'M cyl dia	INPUT VALUE		
8	'Caliper ...			
9	'...1st piston dia	INPUT VALUE	'Piston 1 area	@PI*c9^2/4
10	'...2nd piston dia	INPUT VALUE	'Piston 2 area	@PI*c10^2/4
11	'Total area	INPUT VALUE	*see note 2*	
12				
13	'Pad/disc gap	INPUT VALUE	'Travel at lever *see note 3*	@ROUND (c13*c11*c4/c5/g5,2)
14	'Pad/disc μ	INPUT VALUE	'Force at disc	g4*c11*c14/g5
15	'Disc eff dia	INPUT VALUE	'Force at tyre	g14*c15/c16
16	'Wheel roll dia	INPUT VALUE	'Tyre μ	INPUT VALUE
17	'Wheelbase	INPUT VALUE	'Axle load	@ROUND ((c20 * (c17 –c18) + G15*c19)/c17, 1)
18	'CG from f spindle	INPUT VALUE	'Slide y = 1 n = 0 *see note 4*	@IF(g15 >g16*g17, 1, 0)
19	'CG from ground	INPUT VALUE	'Overturn y = 1 n = 0	@IF(g17*g16 *c19>c20*c18, 1, 0)
20	'Total weight	INPUT VALUE		

Note:

1 @PI is π or 3.14159.
2 The total piston area depends on the caliper type. It will usually be (1st piston area × 2) + (2nd piston area × 2)
3 @ROUND (x, n) means print the value of x to n decimal places.
4 @IF (test, a, b) means if test is *true*, return the value *a* otherwise return the value *b*.

The program shows immediately the full effects of any change, from increasing hand pressure to changing the size of one of the hydraulic cylinders or the wheel itself. It does not show the effects of weight transfer or the amount of suspension compression this would cause, but this can be fed in manually by altering the height of the centre of gravity. In fact, one of the most useful applications of the program is to discover the optimum height of the centre of gravity so that the front suspension and anti-dive can be devised accordingly.

Engine air flow

A 750 four-stroke running at 10,000 rpm will, nominally, shift 3750 litres of air every minute (or 62.5 litres or 110 pints a second). The flow into the air box, or into the area around the engine intakes has to be in this order. If the air box or the bike's bodywork restrict it, then the air pressure at the intake will fall, which may affect the carburation (it will tend to richen it) and the air flow to the engine will fall, so the power will drop. It may be possible to correct the carburation by having the float bowl(s) vented to the same place that the engine gets its air from, but it is obviously desirable to make sure that the required flow is not restricted in the first place.

When a gas travels at speed, it exhibits less pressure energy in order to compensate for its increase in kinetic energy (Bernoulli's theorem). The air travelling past the side of the bike therefore appears to have a pressure which is lower than static air. It encourages flow from the under-tank region *outwards*, which does not help the engine at all.

Air scoops at the front of the bike, in a high pressure region, can deliver air directly to the air box, or to the vicinity of the air box (the former would cause a change in carburation as the speed increases). This can supply the air flow that the engine needs and may be able to increase the air pressure at very high speeds. The best evidence for this is the Ducati 851 Kit which produced 86 bhp as stock and 97 bhp with the air box removed, proving that the air box, or its small, forward facing intakes were restrictive (an easy test to do on the Ducati as its Weber IAW injectors compensated for the changes automatically). With the air box properly in place, the bike would reach 160 mph which, for the size and shape of the Ducati, normally takes about 100 bhp.

The amount of air needed can be worked out from the fuel flow at various engine speeds. If the engine needs a mixture of 12 or 13:1 air:fuel by weight, then the weight of air required is 13 × the fuel flow.

The amount of air scooped depends on the speed and the effective area of the scoop(s) – which will be somewhat less than the geometric area if the pipe is not smooth and straight with a generous radius at its entry (see Chapter 7).

If the effective area is A (ft^2 or m^2), the speed is v (ft/s or m/s) and the air density is d (lbf/ft^3 or kg/m^3) then the weight of air picked up is w (lbf/s or kg/s):

$$w = Avd$$

252

Assuming an air density of 0.08 lbf/ft³ (1.29 kg/m³) and typical air flow rates/power levels for various speeds, then the area of the scoop needed is shown in Table A.2. One column just matches the engine's air flow need at maximum speed, the other matches its need at 100 ft/s. If the scoop is larger than required then it adds to the drag of the bike and is obviously more difficult to install. If the scoop pressurizes the air box at high speed then the carburettors should be vented to the air box and carburation checked at various speeds.

Table A.2 Engine air flow v. road speed

Max speed ft/s	Air flow needed lbf/h	Scoop area to provide flow at max speed in²	Scoop area to provide flow at 100 ft/s in²
176	420	1.18	2.10
198	600	1.50	3.00
220	840	1.90	4.20
242	1020	2.09	5.10
257	1200	2.32	6.00

Thermocouples

In a closed circuit made from two different metals, a small current flows if the two junctions are at different temperatures. This current (and the voltage that produces it) depends on the temperature difference between the junctions and on the two metals. A millivoltmeter connected into one of the materials can be calibrated to read in degrees (usually by immersing the cold junction in a container of water mixed with ice and by immersing the other in water which is heated), using a thermometer or using its boiling point of 100°C.

Figure A.11 Thermocouples. The analogue meter is made by VDO and has the hot junction silver soldered to a washer which fits under a spark plug. The digital instrument, made by Digitron, has plug-in probes, one of which has been joined to a bleed nipple so that it can monitor the temperature of brake fluid inside the caliper

253

Ready-made thermocouples can be bought for certain applications, or with a variety of probes. Copper-constantan thermocouples are used for temperatures up to 600 °C, iron-constantan for temperatures up to 900 °C, Table A.3 shows the thermoelectric potential of various materials, using platinum as a base (the exact values are subject to impurities in the metals).

Table A.3 Thermocouples

The voltage produced in pairs of metals for each °C difference between the two junctions.

Material	Potential (mV/°C) (average values)
Nickel chrome	+0.022
Iron	+0.0188
Steel	+0.0077
Copper	+0.0075
Tin	+0.0044
Lead	+0.0042
Aluminium	+0.0039
Platinum	0
Nickel	−0.016
Constantan	−0.033

The further apart the materials are on the scale, the greater the effect, which explains the popularity of constantan.

Index

Printed and bound by CPI Group (UK) Ltd, Croydon, CR0 4YY

03/10/2024

01040435-0010